STRESS WAVES
IN SOLIDS

BY

H. KOLSKY

Professor of Applied Physics
Brown University

NEW YORK
DOVER PUBLICATIONS, INC.

Published in Canada by General Publishing Com-
pany, Ltd., 30 Lesmill Road, Don Mills, Toronto,
Ontario.
Published in the United Kingdom by Constable
and Company, Ltd., 10 Orange Street, London
WC 2.

This Dover edition, first published in 1963, is an
unabridged and corrected republication of the work
originally published by the Clarendon Press, Oxford,
in 1953, to which has been added a new "List of
books and review articles on subjects discussed in
this monograph which have appeared since 1953,"
especially prepared for this edition by the author.
This edition is published by special arrangement
with Oxford University Press.

Standard Book Number: 486-61098-5
Library of Congress Catalog Card Number: 63-21676

Manufactured in the United States of America
Dover Publications, Inc.
180 Varick Street
New York, N. Y. 10014

PREFACE

THE theory of the propagation of elastic waves in solids was developed during the last century by Stokes, Poisson, Rayleigh, Kelvin, and others as an extension of the theory of elasticity to the problem of vibrating bodies, and also as a help in their studies of the transmission of light considered as vibrations of an elastic ether. During the first quarter of this century the subject was neglected by physicists, partly because of the rival attractions of new fields opened up by the discoveries of atomic physics, and partly because the theory was in many ways in advance of experimental work, as there were then no methods available for observing the passage of stress waves on a laboratory scale. As a result of a number of separate causes, there has in recent years been a remarkable revival of interest in the subject, and a large and increasing number of original papers on both the experimental and the theoretical aspects are now appearing.

The reasons for this change are first that, as a result of the development of electronic techniques, elastic waves of high frequency can be produced and detected easily and ultrasonics is rapidly becoming a subject of its own. Secondly, the development of new materials such as plastics has led to an increased interest in the theory of the mechanical behaviour of imperfectly elastic solids, and stress waves provide a powerful tool for studying the mechanical properties of these substances. Lastly the investigation of the properties of solids at very high rates of loading has become increasingly important from an engineering point of view, and problems associated with the propagation of stress pulses of large amplitude and short duration are not unnaturally matters of considerable military importance. These were studied intensively during the Second World War and have led to the development of the theory of plastic waves.

The purpose of this monograph is to give a concise account of the classical theory, to consider how this theory has been extended to solids which are not perfectly elastic, and then to summarize the experimental work carried out in recent years. The subject necessarily involves some mathematical analysis, but this has been kept as simple as possible and should be followed with ease by physics graduates.

The author wishes to express his thanks to Professor Willis Jackson, Professor N. F. Mott, Dr. R. Hill, Dr. E. G. Stanford, and Mr. S. H. Moss for their valued advice, to many of his colleagues, and in particular to Mr. M. T. Sampson and Mr. D. G. Christie, for reading the manuscript, and finally to Professor R. M. Davies, without whose continued help and encouragement this monograph would not have been completed.

H. K.

BUTTERWICK RESEARCH LABORATORIES
IMPERIAL CHEMICAL INDUSTRIES LIMITED
WELWYN, HERTS.

July 1952

CONTENTS

PART II

STRESS WAVES IN IMPERFECTLY ELASTIC MEDIA

Plates II *and* III *face pp.* 192 *and* 193 *respectively*

PRINCIPAL SYMBOLS USED

a radius of cylinder.

c velocity of propagation of waves in general.

c_0 velocity of longitudinal waves of infinite wavelength in a bar $= (E/\rho)^{\frac{1}{2}}$.

c_1 velocity of dilatation waves in an unbounded medium $= [(\lambda+2\mu)/\rho]^{\frac{1}{2}}$.

c_2 velocity of distortion waves in an unbounded medium $= (\mu/\rho)^{\frac{1}{2}}$.

c_g group velocity of longitudinal waves in a bar.

c_p phase velocity of longitudinal waves in a bar.

c_s velocity of Rayleigh surface waves.

c' phase velocity of flexural waves in a bar.

$c_g{}'$ group velocity of flexural waves in a bar.

$c_{11}, c_{12}, \ldots$ elastic constants in an aeolotropic medium.

E Young's modulus.

f 2π divided by wavelength Λ in the x-direction.

g 2π divided by wavelength Λ in the y-direction.

h $= [\rho p^2/(\lambda+2\mu)]^{\frac{1}{2}}$ for Rayleigh waves.

h' $= [\rho p^2/(\lambda+2\mu)-\gamma^2]^{\frac{1}{2}}$ in Pochhammer treatment.

i $\sqrt{(-1)}$.

I second moment of a cross-section of bar about its diameter.

I' moment of inertia of unit length of cylindrical bar about the axis.

$J_\beta(\ )$ Bessel function of order β.

k bulk modulus.

K radius of gyration of section of bar about diameter.

l, m, n direction cosines.

N resonant frequency.

p 2π times the frequency of sinusoidal waves.

P force, in general; hydrostatic pressure in treatment of shock waves.

Q measure of sharpness of resonance.

$\mathbf{s}$ displacement vector.

S $= d\sigma'/d\epsilon$ in plastic region.

t time.

u, v, w displacements in x, y, and z directions respectively in Cartesian coordinates.

u_r, u_θ, u_z displacements in r, θ, and z directions respectively in cylindrical polar coordinates.

U, V, W amplitudes of vibration in Pochhammer treatment.

V particle velocity, in general.

X Lagrangian coordinate for plastic waves.

α attenuation factor for stress waves in dissipative media.

β $= X/t$ in treatment of plastic waves.

γ 2π divided by wavelength Λ in the z-direction.

δ phase angle.

Δ dilatation $= \epsilon_{xx} + \epsilon_{yy} + \epsilon_{zz}$.

ΔW energy dissipated in stress cycle.

ϵ strain.

$\epsilon_{xx}, \epsilon_{xy}, \epsilon_{xz}$, etc., components of strain in Cartesian coordinates.

κ $= [\rho p^2/\mu]^{\frac{1}{2}}$ for Rayleigh waves.

κ_1 $= \kappa/f$ for Rayleigh waves.

κ' $= [\rho p^2/\mu - \gamma^2]^{\frac{1}{2}}$ in Pochhammer treatment.

Λ wavelength.

λ Lamé's constant.

μ rigidity modulus.

ν Poisson's ratio.

ξ displacement.

ρ density.

σ stress.

σ' engineering stress.

$\sigma_{rr}, \sigma_{r\theta}, \sigma_{rz}$ components of stress along the radius vector r in cylindrical polar coordinates.

$\sigma_{xx}, \sigma_{xy}, \sigma_{xz}, \sigma_{yx}$, etc., components of stress in Cartesian coordinates.

ϕ potential function of displacement for irrotational strain.

$\bar{\omega}_r, \bar{\omega}_\theta, \bar{\omega}_z$ components of rotation in cylindrical polar coordinates.

$\bar{\omega}_x, \bar{\omega}_y, \bar{\omega}_z$ components of rotation in Cartesian coordinates.

∇^2 Laplacian operator

$$\left(= \frac{\partial^2}{\partial x^2} + \frac{\partial^2}{\partial y^2} + \frac{\partial^2}{\partial z^2} \text{ in Cartesian coordinates} \right).$$

CHAPTER I

INTRODUCTION

In rigid dynamics it is assumed that, when a force is applied to any one point on a body, the resultant stresses set every other point in motion instantaneously, and the force can be considered as producing a linear acceleration of the whole body, together with an angular acceleration about its centre of gravity. In the theory of elasticity, on the other hand, the body is considered as in equilibrium under the action of applied forces, and the elastic deformations are assumed to have reached their static values. These treatments are sufficiently accurate for problems in which the time between the application of a force and the setting up of effective equilibrium is short compared with the times in which the observations are made. When, however, we are considering the effects of forces which are applied for only very short periods of time, or are changing rapidly, the effects must be considered in terms of the propagation of stress waves. It is the purpose of this monograph to describe the problems which are associated with the treatment of stress waves in solids, and the methods used to investigate them both experimentally and theoretically.

The finite velocity of waves in a fluid of density ρ and bulk modulus k can be inferred directly from the equation of motion for such a medium, and it is shown in textbooks on the theory of sound that this velocity of propagation will be given by $\sqrt{(k/\rho)}$. When the medium cannot sustain finite shear stresses, this is the only type of wave motion which can be propagated through it. In extended isotropic solids, however, two types of elastic wave may be propagated. These are waves of dilatation which travel with the velocity $[(k+\frac{4}{3}\mu)/\rho]^{\frac{1}{2}}$, μ being the rigidity modulus, and waves of distortion which travel with the velocity $(\mu/\rho)^{\frac{1}{2}}$. When a solid medium is deformed, both distortional and dilatational waves will normally be produced, and it may be shown that when a wave of either type impinges on a boundary of the solid, waves of both types are generated.

In addition to these two types of wave which can travel through an extended solid medium, elastic waves may be propagated along the surface of a solid; these are known as Rayleigh waves, and the disturbances associated with them decay exponentially with depth. Since these waves spread only in two dimensions, they fall off more slowly with distance than the other types of elastic wave. They are of importance in seismic phenomena.

Real solids are never perfectly elastic, so that when a disturbance is propagated through them some of the mechanical energy is converted to heat, and the several different mechanisms by which this takes place have been collectively termed *internal friction*. When a solid is taken through a stress cycle, it generally shows a hysteresis loop; i.e. the stress-strain curve for decreasing stresses does not retrace its upward path exactly. Even when the magnitude of this effect is negligible for static loading, it may be an important factor in the attenuation of stress waves, since in the passage of a pressure pulse through a material each layer in turn is taken through such a cycle, whilst for sinusoidal vibrations the number of hysteresis cycles will depend on the frequency, and may be of the order of millions per second. The velocity gradients set up by the stress wave result in a second type of loss, associated with the viscosity of the material. The nature of the attenuation is different for these two types of internal friction, and the experimental evidence suggests that both types occur.

Many materials also show *mechanical relaxation*; this means that the strain produced by the sudden application of a fixed stress increases asymptotically with time. Similarly, the stress produced when the material is suddenly strained relaxes asymptotically. It is found that stress waves whose periods are close to the *relaxation times* of such a medium are severely attenuated when passing through it. Lastly, the compressions and dilatations caused by stress waves produce temperature gradients, and the finite thermal conductivity of the medium provides another mechanism by which the mechanical energy of the waves may be dissipated as thermal energy.

The nature of internal friction in solids is not at all well understood, and further experimental and theoretical work in this field is required. Its study should lead to a better knowledge of the molecular processes which occur when solids are deformed, and hence of the relation between molecular structure and macroscopic physical properties.

Two other types of stress wave, occurring in media whose stress-strain relation has ceased to be linear, are of importance; viz., shock waves and plastic waves. A shock wave may be formed in a medium when the velocity of propagation of large disturbances in it is greater than that of smaller ones. Under these conditions any pressure pulse, on passing through the medium, develops a steeper and steeper front, and the thickness of this front is ultimately determined by the molecular constitution of the medium. Plastic waves, on the other hand, can be produced when the medium is elastic up to a given stress, but for stresses greater than this, flow occurs. Under these conditions an elastic wave is propagated through the medium, and this is followed by a plastic wave which travels with a lower velocity.

Part I of this monograph treats the propagation of stress waves in perfectly elastic solids, and the theory is developed as a mathematical consequence of Hooke's law and the equations of motion. The only difference between individual solids in this treatment results from differences in the values of their elastic constants and their densities. At the end of Part I recent experimental work concerned with the verification of the theory is described.

Part II is concerned with the propagation of stress waves through solids which are not perfectly elastic. The measurement of internal friction and the nature of the various dissipative processes which cause it are discussed first. A review of experimental work on the measurement of dynamic elastic properties is then given. Finally, the theory of plastic waves and shock waves is outlined and some of the fracture phenomena produced by large stress pulses are described.

PART I

ELASTIC WAVES

CHAPTER II

PROPAGATION IN AN EXTENDED ELASTIC MEDIUM

In this chapter the equations of motion of an isotropic elastic medium will be derived in terms of the particle displacements, and it will be shown that these equations of motion correspond to two types of waves which can be propagated through an extended elastic solid. These two types of wave are termed *dilatational* and *distortional*. The particle motion in a plane dilatational wave is along the direction of propagation, whilst in a plane distortional wave it is perpendicular to the direction of propagation.

If the solid is unbounded these are the only two types of wave which can be propagated through it. When the solid has a free surface or where there is a surface boundary between two solids Rayleigh surface waves may also be propagated; these are discussed later in the chapter. Finally the reflection and refraction of elastic waves at plane boundaries is treated and a short outline of propagation of elastic waves in crystalline media is given.

For the benefit of readers who wish to omit the mathematical analysis involved in Chapters II and III, a summary of the principal results of the theory of elastic waves in extended and bounded solids is given at the end of Chapter III (pp. 84 et seq.).

Components of stress and strain

The stress on a surface element in a solid body does not, in general, act normally to that surface, but has components both normal to the plane and tangential to it. If we refer the body to three mutually perpendicular axes Ox, Oy, Oz, and consider the stresses acting on three planes normal to these axes which pass through a point P, there will be nine components of stress. These will be denoted by σ_{xx}, σ_{yy}, σ_{zz}, σ_{xy}, σ_{xz}, etc., the first letter in the suffix denoting the direction of the stress and the second letter defining the plane in which it is acting. There are several different notations for stress in current use and these are compared in the Appendix. By considering an infinitesimal rectangular parallele-

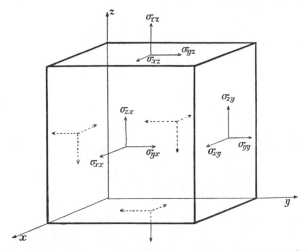

Fig. 1. Stress components acting on an infinitesimal rectangular parallelepiped.

piped round P, with its faces normal to the axes (see Fig. 1) and taking moments, it may be seen that for equilibrium $\sigma_{xy} = \sigma_{yx}$, $\sigma_{xz} = \sigma_{zx}$, and $\sigma_{yz} = \sigma_{zy}$, so that only six independent components of stress remain. From these six components the stresses acting on any other element of surface passing through P may be deduced† and they thus completely define the stress at the point.

The *displacement* of any point P in the body may be resolved parallel to the x, y, and z axes into components u, v, and w, so that, if the coordinates of the point in the undisplaced position

† This may be seen by considering an infinitesimal tetrahedron round P with one of its sides parallel to the element of surface and the other three sides normal to the three coordinate axes. By resolving the forces acting on the tetrahedron in three perpendicular directions, the stresses acting on the inclined element are obtained in terms of the six stress components.

were (x, y, z) they become $(x+u, y+v, z+w)$. In order to define the *strain* at the point we must consider how its position relative to adjacent points has changed. Consider a point very close to P, which in the undisplaced position had coordinates $(x+\delta x)$, $(y+\delta y)$, $(z+\delta z)$, and let the displacement which it has undergone have components $(u+\delta u, v+\delta v, w+\delta w)$, then if δx, δy, and δz are sufficiently small, we have:

$$\delta u = \frac{\partial u}{\partial x}\,\delta x + \frac{\partial u}{\partial y}\,\delta y + \frac{\partial u}{\partial z}\,\delta z,$$

$$\delta v = \frac{\partial v}{\partial x}\,\delta x + \frac{\partial v}{\partial y}\,\delta y + \frac{\partial v}{\partial z}\,\delta z,$$

$$\delta w = \frac{\partial w}{\partial x}\,\delta x + \frac{\partial w}{\partial y}\,\delta y + \frac{\partial w}{\partial z}\,\delta z.$$

Thus, if the values of the nine quantities

$$\frac{\partial u}{\partial x}, \frac{\partial u}{\partial y}, \frac{\partial u}{\partial z}, \frac{\partial v}{\partial x}, \frac{\partial v}{\partial y}, \frac{\partial v}{\partial z}, \frac{\partial w}{\partial x}, \frac{\partial w}{\partial y}, \frac{\partial w}{\partial z}$$

at the point are known, the relative displacement of all surrounding points may be found. For convenience these nine quantities are regrouped and denoted as follows:

$$\left.\begin{aligned}
\epsilon_{xx} &= \frac{\partial u}{\partial x}, & \epsilon_{yy} &= \frac{\partial v}{\partial y}, & \epsilon_{zz} &= \frac{\partial w}{\partial z} \\[2mm]
\epsilon_{yz} &= \frac{\partial w}{\partial y} + \frac{\partial v}{\partial z}, & \epsilon_{zx} &= \frac{\partial u}{\partial z} + \frac{\partial w}{\partial x}, & \epsilon_{xy} &= \frac{\partial v}{\partial x} + \frac{\partial u}{\partial y} \\[2mm]
2\bar{\omega}_x &= \frac{\partial w}{\partial y} - \frac{\partial v}{\partial z}, & 2\bar{\omega}_y &= \frac{\partial u}{\partial z} - \frac{\partial w}{\partial x}, & 2\bar{\omega}_z &= \frac{\partial v}{\partial x} - \frac{\partial u}{\partial y}
\end{aligned}\right\}. \quad (2.1)$$

The first three quantities, ϵ_{xx}, ϵ_{yy}, and ϵ_{zz} may be seen to correspond to the fractional expansions and contractions of infinitesimal line elements passing through P parallel to the x, y, and z axes respectively. The second three, ϵ_{yz}, ϵ_{zx}, and ϵ_{xy} correspond to the components of shear strain in the planes denoted by their suffixes. The last three, $\bar{\omega}_x$, $\bar{\omega}_y$, and $\bar{\omega}_z$ do not correspond to a deformation of the element round P, but are the components of its rotation as a rigid body, and if the displacement is regarded as a vector **s** these will be the components of $\frac{1}{2}$ curl **s** along the three axes (see Appendix).

The meaning of these quantities for a two-dimensional strain in the yz plane is shown in Fig. 2. $ABCD$ is an infinitesimal square which has been displaced and deformed into the rhombus $A'B'C'D'$, θ_1 and θ_2 being the angles $A'D'$ and $A'B'$ make

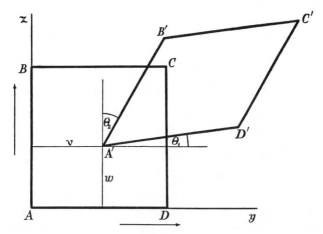

FIG. 2. Shear and rotation in two dimensions.

with the y and z axes respectively. Now $\tan\theta_1 = \partial w/\partial y$, and $\tan\theta_2 = \partial v/\partial z$, and for sufficiently small deformations the angles may be taken equal to their tangents so that $\epsilon_{yz} = \theta_1 + \theta_2$, and this is by definition the shear strain. It may also be seen from the figure that the quantity $2\bar{\omega}_x = \theta_1 - \theta_2$ does not correspond to a deformation of $ABCD$ but to twice the angle through which AC has been rotated. The first six quantities defined in equations (2.1) are called the components of strain. When the last three are zero, the deformation is irrotational and is described as *pure strain*.

Generalized form of Hooke's law

It is found experimentally that for most solids the observed strains are proportional to the applied load, provided that the load does not exceed a given value which is known as the elastic limit. This experimental law is stated mathematically as: *Each of the six components of stress is at any point a linear function of the six components of strain.* Whilst the law in this form is incapable of direct experimental proof, it summarizes the experimental results for different types of loading, and wherever the mathematical consequences can be tested they are found to be true within the elastic range of the material.

This generalized form of Hooke's law may thus be written:

$$\left. \begin{aligned}
\sigma_{xx} &= c_{11}\,\epsilon_{xx} + c_{12}\,\epsilon_{yy} + c_{13}\,\epsilon_{zz} + c_{14}\,\epsilon_{yz} + c_{15}\,\epsilon_{zx} + c_{16}\,\epsilon_{xy} \\
\sigma_{yy} &= c_{21}\,\epsilon_{xx} + c_{22}\,\epsilon_{yy} + c_{23}\,\epsilon_{zz} + c_{24}\,\epsilon_{yz} + c_{25}\,\epsilon_{zx} + c_{26}\,\epsilon_{xy} \\
\sigma_{zz} &= c_{31}\,\epsilon_{xx} + c_{32}\,\epsilon_{yy} + c_{33}\,\epsilon_{zz} + c_{34}\,\epsilon_{yz} + c_{35}\,\epsilon_{zx} + c_{36}\,\epsilon_{xy} \\
\sigma_{yz} &= c_{41}\,\epsilon_{xx} + c_{42}\,\epsilon_{yy} + c_{43}\,\epsilon_{zz} + c_{44}\,\epsilon_{yz} + c_{45}\,\epsilon_{zx} + c_{46}\,\epsilon_{xy} \\
\sigma_{zx} &= c_{51}\,\epsilon_{xx} + c_{52}\,\epsilon_{yy} + c_{53}\,\epsilon_{zz} + c_{54}\,\epsilon_{yz} + c_{55}\,\epsilon_{zx} + c_{56}\,\epsilon_{xy} \\
\sigma_{xy} &= c_{61}\,\epsilon_{xx} + c_{62}\,\epsilon_{yy} + c_{63}\,\epsilon_{zz} + c_{64}\,\epsilon_{yz} + c_{65}\,\epsilon_{zx} + c_{66}\,\epsilon_{xy}
\end{aligned} \right\}, \quad (2.2)$$

where the coefficients are the elastic constants of the material.

It may be shown (e.g. Love, 1927, p. 99) that the condition for the elastic energy to be a univalued function of the strain is that any coefficient c_{rs} is equal to the coefficient c_{sr}. This relation reduces the number of independent coefficients from 36 to 21. In a completely aeolotropic material where no particular spacial symmetry exists (e.g. a triclinic crystal) the values of twenty-one different quantities must be known in order to define the elastic properties of the medium. Where the material has axes or planes of symmetry, relations may be established between these coefficients (see Love, 1927, p. 149), and the number of independent elastic constants is consequently reduced. Thus for a cubic crystal there are only three independent constants.

In an isotropic solid the values of the coefficients must be independent of the set of rectangular axes chosen, and if this condition is applied to the equations only two independent constants remain. These will be denoted by λ and μ. We then have

$$c_{12} = c_{13} = c_{21} = c_{23} = c_{31} = c_{32} = \lambda,$$

$$c_{44} = c_{55} = c_{66} = \mu,$$

$$c_{11} = c_{22} = c_{33} = \lambda + 2\mu,$$

and the other twenty-four coefficients all become zero. Equations (2.2) may then be written:

$$\sigma_{xx} = \lambda\Delta + 2\mu\epsilon_{xx}, \qquad \sigma_{yy} = \lambda\Delta + 2\mu\epsilon_{yy}, \qquad \sigma_{zz} = \lambda\Delta + 2\mu\epsilon_{zz} \left.\begin{array}{c}\\\\\end{array}\right\},$$
$$\sigma_{yz} = \mu\epsilon_{yz}, \qquad \sigma_{zx} = \mu\epsilon_{zx}, \qquad \sigma_{xy} = \mu\epsilon_{xy}$$

$$(2.3)$$

where $\Delta = \epsilon_{xx} + \epsilon_{yy} + \epsilon_{zz}$; this represents the change in volume of a unit cube and is called the dilatation.

The two elastic constants, λ and μ, are known as Lamé's constants, and they completely define the elastic behaviour of an isotropic solid. For convenience, however, four elastic constants are normally used. These are Young's modulus E, Poisson's ratio ν, the bulk modulus k, and the rigidity modulus which is identical with Lamé's constant μ. Using equation (2.3) E, ν, and k may be expressed in terms of λ and μ. E may be defined as the ratio between the applied stress and the fractional extension, when a cylindrical or prismatic specimen is subjected to a uniform stress over its plane ends and its lateral surfaces are free from constraint. If the x-axis is taken parallel to the axis of the cylinder, σ_{xx} is the applied stress and the other five components of stress are zero.

The first three equations thus become:

$$\sigma_{xx} = (\lambda + 2\mu)\epsilon_{xx} + \lambda(\epsilon_{yy} + \epsilon_{zz})$$
$$0 = (\lambda + 2\mu)\epsilon_{yy} + \lambda(\epsilon_{xx} + \epsilon_{zz})$$
$$0 = (\lambda + 2\mu)\epsilon_{zz} + \lambda(\epsilon_{xx} + \epsilon_{yy}),$$

and solving for ϵ_{xx}, ϵ_{yy}, ϵ_{zz}, we have:

$$\epsilon_{xx} = \frac{\lambda + \mu}{\mu(3\lambda + 2\mu)}\sigma_{xx} \quad \text{and} \quad \epsilon_{yy} = \epsilon_{zz} = -\frac{\lambda}{2\mu(3\lambda + 2\mu)}\sigma_{xx}.$$

Young's modulus E is given by $\sigma_{xx}/\epsilon_{xx}$, so that:

$$E = \frac{\mu(3\lambda + 2\mu)}{\lambda + \mu}. \qquad (2.4)$$

Poisson's ratio ν is defined as the ratio between the lateral contraction and the longitudinal extension of the specimen, the lateral surfaces again being free, i.e. $-\epsilon_{yy}/\epsilon_{xx}$. Hence

$$\nu = \frac{\lambda}{2(\lambda + \mu)}. \qquad (2.5)$$

The bulk modulus k is defined as the ratio between the applied pressure and the fractional change in volume when the solid is subjected to uniform hydrostatic compression. Under these conditions

$$\sigma_{xx} = \sigma_{yy} = \sigma_{zz} = -P \text{ (say)}, \quad \text{and} \quad \sigma_{yz} = \sigma_{zx} = \sigma_{xy} = 0,$$

so that, from (2.3), $\epsilon_{xx} = \epsilon_{yy} = \epsilon_{zz} = -P/(3\lambda+2\mu)$.

The fractional change in volume is given by

$$-(\epsilon_{xx}+\epsilon_{yy}+\epsilon_{zz}) = -\Delta,$$

so that:
$$k = \frac{P}{\Delta} = \lambda+\frac{2\mu}{3}. \tag{2.6}$$

Finally the shear modulus or rigidity is μ, and itself corresponds to the ratio between the shear stress and the shear strain as given by the last three equations of (2.3).

Equations of motion in an elastic medium

In order to obtain the equations of motion for an elastic medium we consider the variation in stress across a small parallelepiped with its sides parallel to a set of rectangular axes (Fig. 3). The components of stress will vary across the faces; to obtain the force acting on each face we take the value of stress at the centre of each face times the area of the face.

As can be seen from the figure, six separate forces will be acting parallel to each axis, and if we consider the resultant force acting in the x-direction, we have:

$$\left(\sigma_{xx}+\frac{\partial\sigma_{xx}}{\partial x}\,\delta x\right)\delta y\delta z-\sigma_{xx}\,\delta y\delta z+$$

$$+\left(\sigma_{xy}+\frac{\partial\sigma_{xy}}{\partial y}\,\delta y\right)\delta x\delta z-\sigma_{xy}\,\delta x\delta z+$$

$$+\left(\sigma_{xz}+\frac{\partial\sigma_{xz}}{\partial z}\,\delta z\right)\delta x\delta y-\sigma_{xz}\,\delta x\delta y$$

which simplifies to

$$\left(\frac{\partial\sigma_{xx}}{\partial x}+\frac{\partial\sigma_{xy}}{\partial y}+\frac{\partial\sigma_{xz}}{\partial z}\right)\delta x\delta y\delta z,$$

and by Newton's second law of motion, neglecting body forces such as gravity, this will be equal to $\left[(\rho\,\delta x\delta y\delta z)\dfrac{\partial^2 u}{\partial t^2}\right]$ where ρ

is the density of the element and u is the displacement in the x-direction, so that

$$\rho \frac{\partial^2 u}{\partial t^2} = \frac{\partial \sigma_{xx}}{\partial x} + \frac{\partial \sigma_{xy}}{\partial y} + \frac{\partial \sigma_{xz}}{\partial z};$$

similarly

$$\rho \frac{\partial^2 v}{\partial t^2} = \frac{\partial \sigma_{yx}}{\partial x} + \frac{\partial \sigma_{yy}}{\partial y} + \frac{\partial \sigma_{yz}}{\partial z} \qquad (2.7)$$

and

$$\rho \frac{\partial^2 w}{\partial t^2} = \frac{\partial \sigma_{zx}}{\partial x} + \frac{\partial \sigma_{zy}}{\partial y} + \frac{\partial \sigma_{zz}}{\partial z}.$$

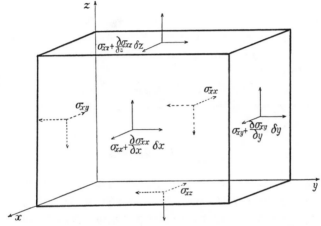

Fig. 3. Stresses acting on a small rectangular parallelepiped.

These equations of motion will hold, whatever the stress-strain behaviour of the medium. In order to solve them, we must use the elastic relations. For an isotropic medium these are given by equations (2.3), and substituting from these for the stress components in (2.7) we have:

$$\rho \frac{\partial^2 u}{\partial t^2} = \frac{\partial}{\partial x}(\lambda \Delta + 2\mu \epsilon_{xx}) + \frac{\partial}{\partial y}(\mu \epsilon_{xy}) + \frac{\partial}{\partial z}(\mu \epsilon_{xz}).$$

Now, by definition, equation (2.1),

$$\epsilon_{xx} = \frac{\partial u}{\partial x}, \qquad \epsilon_{xz} = \frac{\partial w}{\partial x} + \frac{\partial u}{\partial z},$$

and

$$\epsilon_{xy} = \frac{\partial v}{\partial x} + \frac{\partial u}{\partial y}.$$

Hence:
$$\rho \frac{\partial^2 u}{\partial t^2} = (\lambda+\mu) \frac{\partial \Delta}{\partial x} + \mu \nabla^2 u, \tag{2.8}$$

where the operator ∇^2 is written for

$$\left(\frac{\partial^2}{\partial x^2} + \frac{\partial^2}{\partial y^2} + \frac{\partial^2}{\partial z^2}\right);$$

similarly
$$\rho \frac{\partial^2 v}{\partial t^2} = (\lambda+\mu) \frac{\partial \Delta}{\partial y} + \mu \nabla^2 v \tag{2.9}$$

and
$$\rho \frac{\partial^2 w}{\partial t^2} = (\lambda+\mu) \frac{\partial \Delta}{\partial z} + \mu \nabla^2 w. \tag{2.10}$$

These are the equations of motion of an isotropic elastic solid in which body forces are absent, and they may be shown to correspond to the propagation of two types of waves through the medium.

Thus, if we differentiate both sides of equation (2.8) with respect to x, both sides of (2.9) w.r.t. y, and both sides of (2.10) w.r.t. z and add, we have:

$$\rho \frac{\partial^2 \Delta}{\partial t^2} = (\lambda+2\mu) \nabla^2 \Delta. \tag{2.11}$$

This is the wave equation and shows that the dilatation Δ is propagated through the medium with velocity $[(\lambda+2\mu)/\rho]^{\frac{1}{2}}$.

If, on the other hand, we eliminate Δ between (2.9) and (2.10) by differentiating both sides of (2.9) w.r.t. z, and of (2.10) w.r.t. y, and subtract, we have:

$$\rho \frac{\partial^2}{\partial t^2}\left(\frac{\partial w}{\partial y} - \frac{\partial v}{\partial z}\right) = \mu \nabla^2\left(\frac{\partial w}{\partial y} - \frac{\partial v}{\partial z}\right)$$

or
$$\rho \frac{\partial^2 \bar{\omega}_x}{\partial t^2} = \mu \nabla^2 \bar{\omega}_x, \tag{2.12}$$

where $\bar{\omega}_x$ is the rotation about the x-axis as given by equation (2.1). Similar equations may be obtained for $\bar{\omega}_y$ and $\bar{\omega}_z$. Thus the rotation is propagated with velocity $(\mu/\rho)^{\frac{1}{2}}$.

If the dilatation Δ is zero, equation (2.8) becomes

$$\rho \frac{\partial^2 u}{\partial t^2} = \mu \nabla^2 u, \tag{2.13}$$

with similar equations for v and w.

The condition that the rotations $\bar{\omega}_x$, $\bar{\omega}_y$, and $\bar{\omega}_z$ should all vanish is satisfied if u, v, and w satisfy the following conditions:

$$u = \frac{\partial \phi}{\partial x}, \qquad v = \frac{\partial \phi}{\partial y}, \quad \text{and} \quad w = \frac{\partial \phi}{\partial z},$$

where ϕ is a potential function.

Thus $\qquad\qquad \Delta = \nabla^2\phi \quad \text{and} \quad \dfrac{\partial \Delta}{\partial x} = \nabla^2 u.$

Substituting in (2.8), we have:

$$\rho\, \frac{\partial^2 u}{\partial t^2} = (\lambda + 2\mu)\nabla^2 u \tag{2.14}$$

and similarly for v and w.

We thus see that in the interior of an elastic solid waves may be propagated with two different velocities. Waves involving no rotation travel with velocity $[(\lambda + 2\mu)/\rho]^{\frac{1}{2}} = c_1$ (say), whilst waves involving no dilatation travel with velocity $(\mu/\rho)^{\frac{1}{2}} = c_2$ (say). Strictly, these two types of waves should be termed *irrotational* and *equivoluminal* respectively, but, since these names are rather cumbrous, the terms dilatation waves and distortion waves are often used. To some extent the term distortion wave is misleading, since, whilst equivoluminal waves involve distortion without dilatation, irrotational waves involve both. For brevity, however, the term 'distortion wave' will be retained here.

It may be shown that any plane wave propagated through an isotropic elastic medium must travel with one or other of the above velocities. Thus, consider a plane wave propagated parallel to the x-axis. (Since the medium is isotropic, there is no loss in generality in assuming this.) Let its velocity of propagation be c; then the displacements u, v, and w will be functions of a single parameter $\psi = x - ct$.

We then have:

$$\frac{\partial^2 u}{\partial t^2} = c^2 \frac{\partial^2 u}{\partial \psi^2}, \qquad \frac{\partial^2 v}{\partial t^2} = c^2 \frac{\partial^2 v}{\partial \psi^2}, \qquad \frac{\partial^2 w}{\partial t^2} = c^2 \frac{\partial^2 w}{\partial \psi^2},$$

$$\frac{\partial^2 u}{\partial x^2} = \frac{\partial^2 u}{\partial \psi^2}, \qquad \frac{\partial^2 v}{\partial x^2} = \frac{\partial^2 v}{\partial \psi^2}, \qquad \frac{\partial^2 w}{\partial x^2} = \frac{\partial^2 w}{\partial \psi^2},$$

and the differential coefficients with respect to y and z are all zero. If we denote the second differential coefficients of u, v, and w with respect to ψ by u'', v'', and w'', and substitute in the first equation of motion (2.8), we have:

$$\rho c^2 u'' = (\lambda + 2\mu) u'', \tag{2.15}$$

similarly from (2.9) and (2.10) we obtain:

$$\rho c^2 v'' = \mu v'' \tag{2.16}$$

and $$\rho c^2 w'' = \mu w''. \tag{2.17}$$

Equations (2.15), (2.16), and (2.17) can only be satisfied in one of two ways; either $c^2 = (\lambda + 2\mu)/\rho$ and v'' and w'' are zero, or $c^2 = \mu/\rho$, and $u'' = 0$. In the former case we have longitudinal waves, in which the motion is along the direction of propagation, and in the latter the motion is transverse and parallel to the wave front.

The theory of transverse elastic body waves was first investigated by Navier (1821), and, a little later, more rigorously, by Poisson (1827). These treatments appeared about the same time as Fresnel's theory of the transverse nature of light vibrations. Since, prior to this, the concept of transverse vibrations transmitted *through* a medium had not been considered at all, subsequent developments in the theory of elastic waves tended to become associated with discussions on the propagation of light. (See, for example, Stokes, 1848, and Kelvin, 1904.)

The velocity of waves of distortion depends only on the density and shear modulus of the medium, and it might appear at first sight that the velocity of waves of dilatation should depend only on the density and the bulk modulus k. However, $k = \lambda + \frac{2}{3}\mu$, see equation (2.6), so that the velocity of dilatation waves is $[(k + \frac{4}{3}\mu)/\rho]^{\frac{1}{2}}$ and thus the shear modulus as well as the bulk modulus is involved. The physical reason for this is that in the propagation of waves of dilatation the medium is not subjected to a simple compression, but to a combination of compression and shear. For consider a small cube of material in the path of a plane wave of dilatation travelling in the direction of the x-axis; its cross-sectional area normal to the x-direction will not alter during the passage of the wave, whilst its x-dimension will be changed. There is thus a change in the *shape* of the element as well as in its volume, and the resistance of the medium to shear as well as its compressibility comes into play.

Integration of the wave equation

Equations (2.11), (2.12), and (2.13) are all of the form:

$$\frac{\partial^2 \alpha}{\partial t^2} = c^2 \nabla^2 \alpha \qquad (2.18)$$

and when the deformation is a function of only one coordinate, e.g. x, the equation becomes:

$$\frac{\partial^2 \alpha}{\partial t^2} = c^2 \frac{\partial^2 \alpha}{\partial x^2}. \qquad (2.19)$$

The general solution of this is:

$$\alpha = f(x-ct) + F(x+ct),$$

where f and F are arbitrary functions depending on initial conditions, f corresponding to a plane wave travelling along the positive direction of the x-axis, and F to one in the direction opposite to this. For each wave it may be seen that if, at any time t_1, α is a given function of x, i.e., if the deformation has a given shape in the medium, at a later time, t_2, it will have the same shape a distance $c(t_2-t_1)$ away, along the x-axis.

If a disturbance is spreading from a point, the deformation will depend only on the value of r, the radius vector from the point. Since $r^2 = x^2+y^2+z^2$ we have:

$$\frac{\partial^2 \alpha}{\partial x^2} = \frac{x^2}{r^2}\frac{\partial^2 \alpha}{\partial r^2} + \frac{1}{r}\left(\frac{r^2-x^2}{r^2}\right)\frac{\partial \alpha}{\partial r}$$

with similar equations for $\partial^2 \alpha / \partial y^2$ and $\partial^2 \alpha / \partial z^2$, so that equation (2.18) becomes:

$$\frac{\partial^2 \alpha}{\partial t^2} = c^2\left(\frac{\partial^2 \alpha}{\partial r^2} + \frac{2}{r}\frac{\partial \alpha}{\partial r}\right) \quad \text{or} \quad \frac{\partial^2(r\alpha)}{\partial t^2} = c^2\frac{\partial^2(r\alpha)}{\partial r^2}.$$

This is of the same form as equation (2.19), and its solution is:

$$r\alpha = f(r-ct) + F(r+ct), \qquad (2.20)$$

where $f(\ )$ represents a spherical wave diverging from the origin whilst $F(\ )$ represents a converging spherical wave. The amplitude in both cases is inversely proportional to the distance r.

Rayleigh waves

In an unbounded isotropic solid two and only two types of elastic wave can be propagated. Where there is a bounding surface, however, elastic surface waves may also occur. These waves, which are similar to gravitational surface waves in liquids, were first investigated by Lord Rayleigh (1887), who showed that their effect decreases rapidly with depth, and that their velocity of propagation is smaller than that of body waves.

We shall consider the propagation of a plane wave through an elastic medium with a plane boundary, and attempt to find a solution of the equations of motion (2.8), (2.9), and (2.10) which corresponds to a disturbance which is largely confined to the neighbourhood of the boundary and which satisfies the condition that the boundary is free from stress. For simplicity we take the boundary to be the xy plane with z positive towards the interior of the solid, and take the plane wave as travelling in the x-direction. Since the displacements will then be independent of y we may define two potential functions ϕ and ψ such that:

$$u = \frac{\partial \phi}{\partial x} + \frac{\partial \psi}{\partial z} \quad \text{and} \quad w = \frac{\partial \phi}{\partial z} - \frac{\partial \psi}{\partial x}. \tag{2.21}$$

Now, from (2.21), the dilatation Δ will be given by:

$$\Delta = \frac{\partial u}{\partial x} + \frac{\partial w}{\partial z} = \nabla^2 \phi,$$

whilst $\bar{\omega}$, the rotation in the xz plane, is given by:

$$2\bar{\omega}_y = \frac{\partial u}{\partial z} - \frac{\partial w}{\partial x} = \nabla^2 \psi.$$

(These equations show that ϕ is associated with the dilatation produced by the disturbance whilst ψ is associated with the rotation, and the introduction of these two potential functions has thus enabled us to separate the effects of dilatation and rotation in the medium.)

Now the three equations of motion (2.8), (2.9), and (2.10) may be written:

$$\rho \frac{\partial^2}{\partial t^2}(u, v, w) = (\lambda+\mu)\left(\frac{\partial\Delta}{\partial x}, \frac{\partial\Delta}{\partial y}, \frac{\partial\Delta}{\partial z}\right)+\mu\nabla^2(u, v, w)$$

and if the displacements are independent of y the first and third of these equations become, on substituting from (2.21)

$$\rho \frac{\partial}{\partial x}\left(\frac{\partial^2\phi}{\partial t^2}\right)+\rho \frac{\partial}{\partial z}\left(\frac{\partial^2\psi}{\partial t^2}\right) = (\lambda+2\mu) \frac{\partial}{\partial x}(\nabla^2\phi)+\mu \frac{\partial}{\partial z}(\nabla^2\psi)$$

and

$$\rho \frac{\partial}{\partial z}\left(\frac{\partial^2\phi}{\partial t^2}\right)-\rho \frac{\partial}{\partial x}\left(\frac{\partial^2\psi}{\partial t^2}\right) = (\lambda+2\mu) \frac{\partial}{\partial z}(\nabla^2\phi)-\mu \frac{\partial}{\partial x}(\nabla^2\psi).$$

These two equations will be satisfied if:

$$\frac{\partial^2\phi}{\partial t^2} = [(\lambda+2\mu)/\rho]\nabla^2\phi = c_1^2\nabla^2\phi \tag{2.22}$$

and

$$\frac{\partial^2\psi}{\partial t^2} = [\mu/\rho]\nabla^2\psi = c_2^2\nabla^2\psi. \tag{2.23}$$

Now, if we consider a sinusoidal wave of frequency $p/2\pi$ propagated in the x-direction with velocity c and wavelength $2\pi/f$ so that $c = p/f$, we may try as solutions of (2.22) and (2.23):

$$\phi = F(z)\exp[i(pt-fx)] \tag{2.24}$$

and

$$\psi = G(z)\exp[i(pt-fx)], \tag{2.25}$$

where $i = \sqrt{(-1)}$ and F and G are functions which determine the way in which the amplitude of the waves changes with z. Substituting the expression for ϕ from (2.24) in equation (2.22) we obtain:

$$-\frac{p^2}{c_1^2} F(z) = -f^2 F(z)+F''(z),$$

where $F''(z)$ is $(d^2/dz^2)F(z)$. This equation may be written:

$$F''(z)-(f^2-h^2)F(z) = 0, \tag{2.26}$$

where $h = p/c_1$.

The general solution of (2.26) is:

$$F(z) = A\exp(-qz)+A'\exp(qz), \tag{2.27}$$

where

$$q^2 = f^2-h^2. \tag{2.28}$$

The second term in equation (2.27) corresponds to a disturbance which increases with increasing z, and for the type of wave we are here considering A' must be zero.

Similarly, if we substitute the expression for ψ from (2.25) in equation (2.23), we obtain:

$$-\kappa^2 G(z) = -f^2 G(z) + G''(z), \quad \text{where } \kappa = \frac{p}{c_2},$$

and the relevant solution of this is:

$$G(z) = B\exp(-sz), \tag{2.29}$$

where
$$s^2 = f^2 - \kappa^2. \tag{2.30}$$

Thus (2.24) and (2.25) become:

$$\phi = A\exp[-qz + i(pt - fx)]$$

and
$$\psi = B\exp[-sz + i(pt - fx)]. \tag{2.31}$$

We now insert these expressions for ϕ and ψ in the boundary condition, namely that the stress components σ_{zz}, σ_{zy}, and σ_{zx} vanish at the surface where $z = 0$.

Now
$$\sigma_{zz} = \lambda\Delta + 2\mu\,\frac{\partial w}{\partial z}$$

and from (2.21) this becomes, in terms of ϕ and ψ,

$$\sigma_{zz} = (\lambda + 2\mu)\,\frac{\partial^2 \phi}{\partial z^2} + \lambda\,\frac{\partial^2 \phi}{\partial x^2} - 2\mu\,\frac{\partial^2 \psi}{\partial x \partial z}.$$

Substituting from (2.31) this gives at $z = 0$

$$A[(\lambda + 2\mu)q^2 - \lambda f^2] - 2B\mu isf = 0, \tag{2.32}$$

whilst
$$\sigma_{zx} = \mu\!\left(\frac{\partial u}{\partial z} + \frac{\partial w}{\partial x}\right),$$

and so from (2.21) we have:

$$\sigma_{zx} = \mu\!\left(2\,\frac{\partial^2 \phi}{\partial x \partial z} - \frac{\partial^2 \psi}{\partial x^2} + \frac{\partial^2 \psi}{\partial z^2}\right).$$

Substituting for ϕ and ψ from (2.31), we obtain at $z = 0$:

$$2iqfA + (s^2 + f^2)B = 0. \tag{2.33}$$

Eliminating the ratio A/B from (2.32) and (2.33) we find:

$$4\mu qsf^2 = [(\lambda + 2\mu)q^2 - \lambda f^2](s^2 + f^2). \tag{2.34}$$

On squaring both sides of (2.34) and substituting for q and s from (2.28) and (2.30), we have:

$$16\mu^2(f^2-h^2)(f^2-\kappa^2)f^4 = [-(\lambda+2\mu)h^2+2\mu f^2]^2(2f^2-\kappa^2)^2.$$

Dividing both sides of this equation by $\mu^2 f^8$ gives:

$$16(1-h^2/f^2)(1-\kappa^2/f^2) = [2-(\lambda+2\mu)\mu^{-1}h^2/f^2]^2(2-\kappa^2/f^2)^2.$$
$$(2.35)$$

Now
$$h = \frac{p}{c_1} \quad \text{and} \quad \kappa = \frac{p}{c_2}$$

so that $h^2/\kappa^2 = \mu/(\lambda+2\mu)$ and from equation (2.5) it may be seen that this may be expressed purely in terms of Poisson's ratio ν. Thus

$$\frac{\mu}{\lambda+2\mu} = \frac{1-2\nu}{2-2\nu} = \alpha_1^2 \text{ (say)},$$

so that
$$h = \alpha_1 \kappa.$$

Thus, substituting for h and for $(\lambda+2\mu)/\mu$, equation (2.35) becomes:
$$16(1-\alpha_1^2\kappa^2/f^2)(1-\kappa^2/f^2) = (2-\kappa^2/f^2)^4. \qquad (2.36)$$

Multiplying out and putting κ_1 for κ/f, equation (2.36) simplifies to
$$\kappa_1^6-8\kappa_1^4+(24-16\alpha_1^2)\kappa_1^2+(16\alpha_1^2-16) = 0. \qquad (2.37)$$

This is a cubic in κ_1^2, and if the value of ν for the medium is known the equation may be solved numerically. Now $\kappa_1 = \kappa/f = p/fc_2$, p/f is the velocity of the surface waves, and c_2 is the velocity of distortion waves; thus κ_1 gives the ratio between the velocity of surface waves and of waves of distortion. The velocity of propagation of surface waves is thus independent of the frequency $p/2\pi$ and depends only on the elastic constants of the material. There is thus no dispersion of these waves and a plane surface wave will travel without change in form.

The rate at which the waves decay with the depth z depends on the values of the attenuation factors q and s. Equations (2.28) and (2.30) give
$$\frac{q^2}{f^2} = 1-\alpha_1^2\kappa_1^2 \qquad (2.38)$$

and
$$\frac{s^2}{f^2} = 1-\kappa_1^2.$$

Thus from the value of κ_1 the values of q/f and s/f may be obtained. Now from equations (2.21) and (2.31) we have:

$$\left.\begin{aligned}
u &= \frac{\partial\phi}{\partial x}+\frac{\partial\psi}{\partial z} = -(Aife^{-qz}+Bse^{-sz})\exp[i(pt-fx)] \\
w &= \frac{\partial\phi}{\partial z}-\frac{\partial\psi}{\partial x} = -(Aqe^{-qz}-Bife^{-sz})\exp[i(pt-fx)]
\end{aligned}\right\}. \quad (2.39)$$

Substituting for B from (2.33) and taking the real parts of (2.39) we have:

$$\left.\begin{aligned}
u &= Af[e^{-qz}-2qs(s^2+f^2)^{-1}e^{-sz}]\sin(pt-fx) \\
w &= Aq[e^{-qz}-2f^2(s^2+f^2)^{-1}e^{-sz}]\cos(pt-fx)
\end{aligned}\right\}. \quad (2.40)$$

If we now take for example $\nu = \frac{1}{4}$, then $\alpha_1^2 = \frac{1}{3}$ and equation (2.37) becomes:

$$3\kappa_1^6-24\kappa_1^4+56\kappa_1^2-32 = 0$$

or $$(\kappa_1^2-4)(3\kappa_1^4-12\kappa_1^2+8) = 0.$$

Thus $$\kappa_1^2 = 4, \quad 2+\frac{2}{\sqrt{3}}, \quad \text{or} \quad 2-\frac{2}{\sqrt{3}}.$$

From (2.38) the first two of these values for κ_1 may be seen to lead to imaginary values for q/f and s/f and do not therefore correspond to waves of the type postulated. The third value gives $\kappa_1 = 0.9194$, and surface waves thus travel with this fraction of the velocity of distortion waves when $\nu = 0.25$.

Inserting the value of κ_1 in (2.38) gives $q/f = 0.8475$, and $s/f = 0.3933$. From (2.40) the rate at which the amplitude of the displacement along the direction of propagation is attenuated with depth depends on the factor:

$$e^{-qz}-2qs(s^2+f^2)^{-1}e^{-sz}.$$

Substituting the numerical values of q/f and s/f this becomes

$$\exp(-0.8475fz)-0.5773\exp(-0.3933fz).$$

This decreases rapidly with increasing values of fz and becomes zero for $fz = 1.210$. For this value of fz, u is zero for all values of x and t; there is thus a plane at a depth of $1.21/f$ in which there is no motion parallel to the surface. f is by definition 2π divided by the wavelength, so that this corresponds to a depth of 0.193 wavelengths. For greater depths the amplitude once again becomes finite and is of the opposite sign so that the vibrations then take place in the opposite phase.

The rate at which the amplitude of the motion in a direction normal to the surface is attenuated with depth, may be seen from the expression for w in equation (2.40) to depend on the factor:

$$e^{-qz} - 2f^2(s^2+f^2)^{-1}e^{-sz}.$$

Substituting the numerical values in this gives

$$\exp(-0.8475fz) - 1.7321\exp(-0.3933fz).$$

This factor does not change sign, so that there is no finite depth at which the motion in a direction normal to the surface vanishes. As z increases the amplitude of the vibration first increases, reaches a maximum at a depth of 0.076 wavelengths, and then decreases monotonically. At a depth of one wavelength $fz = 2\pi$ and the amplitude has fallen to 0.19 of its value at the surface.

It may be seen that fz is the relevant parameter in the attenuation with depth of the vibrations both parallel to the surface and normal to it. Since p/f is the velocity of propagation of surface waves, which is a constant for any one material, and $p/2\pi$ is the frequency of the vibrations, f is proportional to frequency. Thus Rayleigh waves of high frequency will be attenuated more rapidly with depth than those of low frequency, and the behaviour is analogous to the *skin effect* in the transmission of high-frequency alternating electric currents in conductors.

The expressions for u and w in equation (2.40) show that the path of any particle in the medium is an ellipse, and it is found that its major axis is normal to the surface. For particles at the surface (where $z = 0$) the ratio between the major and minor axes of the ellipse is 1.468.

All the numerical calculations have here been carried out on the assumption that $\nu = \frac{1}{4}$. Similar results are, however, obtained if a different value of ν is used. If, for example, we take the extreme value of $\nu = \frac{1}{2}$, $\alpha_1 = 0$, and equation (2.37) becomes:

$$\kappa_1^6 - 8\kappa_1^4 + 24\kappa_1^2 - 16 = 0$$

which has $\kappa_1^2 = 0.9127$ as its only real root. Hence the surface waves here travel with 0.9554 times the velocity of distortion waves, and substitution in (2.40) shows that motion parallel to the surface vanishes in this case at a depth of 0·138 wavelengths.

For steel $\nu = 0·29$, and when the cubic (2.37) is solved for this value of ν, the velocity of Rayleigh waves comes out as 0·9258 of the velocity of waves of distortion in the material. The author is indebted to Professor R. M. Davies for Fig. 4, which shows

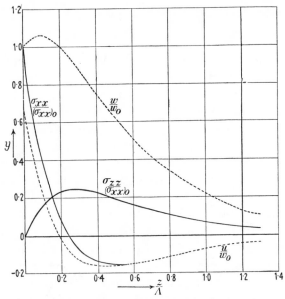

FIG. 4. Amplitudes of the stresses and displacements associated with Rayleigh surface waves in steel. ($\nu = 0·29$.)

the calculated values of the amplitudes of the stress components and displacements for this value of Poisson's ratio. The curves are given in non-dimensional form, the peak values of the displacements $\hat{u}$ and $\hat{w}$ being plotted as the ratios $\hat{u}/\hat{w}_0$ and $\hat{w}/\hat{w}_0$, where $\hat{w}_0$ is the amplitude of the vibrations in the z-direction at the surface. The peak values $\hat{\sigma}_{xx}$, $\hat{\sigma}_{zz}$ of the stresses are plotted as the ratios $\hat{\sigma}_{xx}/(\hat{\sigma}_{xx})_0$ and $\hat{\sigma}_{zz}/(\hat{\sigma}_{xx})_0$ where $(\hat{\sigma}_{xx})_0$ is the amplitude of σ_{xx} at the surface. These ratios are plotted against z/Λ, where Λ is the wavelength of the vibrations and is equal to $2\pi/f$. The curves illustrate how the amplitude of the vibration in the x-direction passes through zero whilst the amplitude in the

z-direction first increases slightly and then decays monotonically. They also show that the σ_{xx} component changes sign, whilst the σ_{zz} component reaches a maximum at about $0 \cdot 3z/\Lambda$ and then decays asymptotically with depth.

Lord Rayleigh suggested that since these surface waves spread only in two dimensions and consequently fall off more slowly with distance than elastic body waves, they might be expected to be of importance in seismic phenomena. This is, in fact, largely borne out by the seismographic records of the waves observed some distance away from an earthquake. These records show three separate groups of waves. The first to arrive are waves in which the vibrations are predominantly longitudinal, these being dilatation waves which have the highest propagation velocity. Next come the distortion waves, in which the motion is found to be mainly transverse, and the third group is of surface waves whose amplitude is large compared with that of the other two. If this last group consisted of pure Rayleigh waves, it should have both vertical and horizontal components, the former predominating. This is not in practice found to be the case, the vertical component sometimes being completely absent. For Rayleigh waves the direction of vibration of the horizontal components should be parallel to the direction of propagation, whereas horizontal components parallel to the wave front are often found. Love (1911) has suggested that these waves can be accounted for by assuming that the elasticity and density of the outer layer of the earth differs from that in the interior. He showed that transverse waves can be propagated through such an outer layer without penetrating into the interior. Waves of this type have become known as *Love waves*.

Stoneley (1924) has considered the more general problem of elastic waves at the surface of separation of two solid media. He has shown that waves analogous to Rayleigh waves will be propagated in the media, the amplitudes in both cases being a maximum at the surface of separation. He has also investigated a generalized type of Love wave which is propagated along an internal stratum bounded on both sides by deep layers of material which differ from it in elastic properties.

Reflection of elastic waves at a free boundary

It has been shown in the earlier sections that two types of elastic wave may be propagated through a solid medium. It is found that, when a wave of either type impinges on a boundary between two media, both reflection and refraction take place. In the most general case, four separate waves are generated; a wave of each type is reflected, and a wave of each type is refracted.

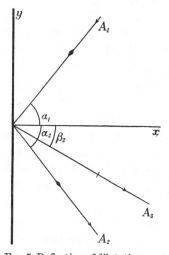

We shall first show that, when a plane wave of dilatation is reflected at a free surface (ideally a surface in vacuum where there can be no refracted waves), the boundary conditions cannot be satisfied by assuming that only a wave of dilatation is reflected. We shall then proceed to find the amplitude and direction of an additional reflected distortion wave which is necessary in order to satisfy these conditions. We shall take the direction of propagation of the incident dilatation wave to be in the xy plane making an angle α_1 with the x-axis, and the free boundary to be the yz plane (see Fig. 5).

FIG. 5. Reflection of dilatation wave at free boundary. (Solid diamonds denote dilatational waves, small bars denote distortional waves.)

If we consider a simple harmonic wave in which the displacement normal to the wave front is denoted by Φ_1 then we may take:

$$\Phi_1 = A_1 \sin(pt + f_1 x + g_1 y),$$

where A_1 is the amplitude of the wave,

$$f_1 = \frac{p \cos \alpha_1}{c_1} \quad \text{and} \quad g_1 = \frac{p \sin \alpha_1}{c_1}$$

(c_1 being the velocity of propagation of the waves). The wave is here taken as travelling in the direction of decreasing x and y.

If u_1 and v_1 are the displacements for this wave parallel to the x and y directions respectively, we have

$$u_1 = \Phi_1 \cos\alpha_1 \quad \text{and} \quad v_1 = \Phi_1 \sin\alpha_1.$$

Now, if a dilatation wave is reflected at an angle α_2 to the x-axis, and its displacement normal to the wavefront is Φ_2, we have:

$$\Phi_2 = A_2 \sin(pt - f_2 x + g_2 y + \delta_1),$$

where $$f_2 = \frac{p\cos\alpha_2}{c_1}, \qquad g_2 = \frac{p\sin\alpha_2}{c_1},$$

δ_1 is a constant to allow for any change in phase of the wave on reflection, and A_2 is the amplitude. If u_2 and v_2 are the displacements produced by the reflected wave,

$$u_2 = -\Phi_2 \cos\alpha_2 \quad \text{and} \quad v_2 = \Phi_2 \sin\alpha_2.$$

Now at the free boundary ($x = 0$) σ_{xx} and σ_{yx} must be zero for all values of y and t, and if we call u and v the net displacements produced by the incident and reflected waves, we have from equation (2.3)

$$\sigma_{xx} = \lambda\Delta + 2\mu\frac{\partial u}{\partial x}$$

and $$\sigma_{yx} = \mu\left(\frac{\partial v}{\partial x} + \frac{\partial u}{\partial y}\right).$$

(Since it has been assumed that there is no displacement in the z-direction, Δ will be given by $[\partial u/\partial x + \partial v/\partial y]$.)

If we substitute $(u_1 + u_2)$ for u, and $(v_1 + v_2)$ for v, we have (after differentiation and rearrangement of the terms):

$$\sigma_{xx} = [\lambda(f_1\cos\alpha_1 + g_1\sin\alpha_1) + 2\mu f_1\cos\alpha_1]\Phi_1' +$$
$$+ [\lambda(f_2\cos\alpha_2 + g_2\sin\alpha_2) + 2\mu f_2\cos\alpha_2]\Phi_2',$$

where $$\Phi_1' = A_1\cos(pt + f_1 x + g_1 y)$$

and $$\Phi_2' = A_2\cos(pt - f_2 x + g_2 y + \delta_1).$$

Substituting for f_1, f_2, g_1, and g_2, this simplifies to

$$\sigma_{xx} = \frac{p}{c_1}[(\lambda + 2\mu\cos^2\alpha_1)\Phi_1' + (\lambda + 2\mu\cos^2\alpha_2)\Phi_2']$$

and at the boundary where $x = 0$, σ_{xx} vanishes, so that we have on substituting for Φ_1' and Φ_2',

$$A_1(\lambda + 2\mu \cos^2 \alpha_1)\cos(pt + g_1 y) +$$
$$+ A_2(\lambda + 2\mu \cos^2\alpha_2)\cos(pt + g_2 y + \delta_1) = 0.$$

This equation can only be satisfied for all values of y and t if $g_1 = g_2$ (i.e. $\alpha_1 = \alpha_2$) and either $\delta_1 = 0$ and $A_1 = -A_2$, or $\delta_1 = \pi$ and $A_1 = A_2$. These two solutions are of course equivalent and correspond to a change in phase of π on reflection.

If we now consider the second condition, namely, that there should be no shear stress on the boundary, and in the same way substitute for u and v in the expression for σ_{yx}, we have:

$$\sigma_{yx} = \mu\left[\frac{\partial}{\partial x}(\Phi_1 \sin \alpha_1 + \Phi_2 \sin \alpha_2) + \frac{\partial}{\partial y}(\Phi_1 \cos \alpha_1 - \Phi_2 \cos \alpha_2)\right]$$

and on differentiation this gives at the boundary ($x = 0$):

$$\sigma_{yx} = \frac{p\mu}{c_1}[A_1 \sin 2\alpha_1 \cos(pt + g_1 y) - A_2 \sin 2\alpha_2 \cos(pt + g_2 y)].$$

This is not identically zero for the conditions required for σ_{xx} to be zero (i.e. $\alpha_1 = \alpha_2$ and $A_1 = -A_2$); thus, if we have only a reflected dilatation wave, we cannot satisfy both boundary conditions, namely, freedom from shear stress and from normal stress. If, however, we assume that, in addition, a wave of distortion is reflected, both boundary conditions can be satisfied. Let the direction of the reflected distortion wave make an angle β_2 with the normal (see Fig. 5), and let the displacement produced by it be Φ_3 then

$$\Phi_3 = A_3 \sin(pt - f_3 x + g_3 y + \delta_2),$$

where $\qquad f_3 = \dfrac{p \cos \beta_2}{c_2}$ and $\quad g_3 = \dfrac{p \sin \beta_2}{c_2}.$

c_2 is the velocity of propagation of distortion waves and δ_2 allows for any phase change on reflection.

The vibrations of this shear wave will be transverse, and, since we assume no motion in the z-direction, the vibrations must take place in the xy plane. If we call the contributions of this wave to the u and v displacements u_3 and v_3, we then have:

$$u_3 = \Phi_3 \sin \beta_2 \quad \text{and} \quad v_3 = \Phi_3 \cos \beta_2.$$

Now let us take first the condition that the shear stress σ_{yx} should be zero at the boundary ($x = 0$); this means

$$\frac{\partial u}{\partial y} + \frac{\partial v}{\partial x} = 0,$$

where u is now $u_1 + u_2 + u_3$ and v is $v_1 + v_2 + v_3$. Substituting for u and v, and differentiating, we have

$$(f_1 \sin \alpha_1 + g_1 \cos \alpha_1)\Phi_1' - (f_2 \sin \alpha_2 + g_2 \cos \alpha_2)\Phi_2' -$$
$$- (f_3 \cos \beta_2 - g_3 \sin \beta_2)\Phi_3' = 0.$$

Substituting for f_1, f_2, etc., and inserting the values of Φ_1', Φ_2' and Φ_3', at $x = 0$, this gives:

$$\frac{A_1}{c_1} p \sin 2\alpha_1 \cos(pt + g_1 y) - \frac{A_2}{c_1} p \sin 2\alpha_2 \cos(pt + g_2 y + \delta_1) -$$
$$- \frac{A_3}{c_2} p \cos 2\beta_2 \cos(pt + g_3 y + \delta_2) = 0.$$

This can only be satisfied for all values of y and t by putting $g_1 = g_2 = g_3$, and hence:

$$\frac{\sin \alpha_1}{c_1} = \frac{\sin \alpha_2}{c_1} = \frac{\sin \beta_2}{c_2}.$$

Hence: $\qquad \alpha_1 = \alpha_2 \quad \text{and} \quad \dfrac{\sin \alpha_1}{\sin \beta_2} = \dfrac{c_1}{c_2}.$

Thus a dilatation wave is reflected at the angle of incidence, and a distortion wave is reflected at an angle similar to that for the refraction of light, the 'refractive index' being the ratio between the velocities of dilatation and of distortion waves, and therefore given by $\sqrt{(2 + \lambda/\mu)}$. We must also have δ_1 and δ_2 equal to 0 or π, and, taking the zero value, the relation between the amplitudes becomes, on substituting for c_1/c_2:

$$2(A_1 - A_2)\cos \alpha_1 \sin \beta_2 - A_3 \cos 2\beta_2 = 0. \qquad (2.41)$$

We can now see whether the condition of absence of normal stress at the boundary can also be satisfied. We have:

$$\sigma_{xx} = (\lambda + 2\mu)\frac{\partial u}{\partial x} + \lambda \frac{\partial v}{\partial y}$$

and this gives at $x = 0$ on substitution of $u = u_1 + u_2 + u_3$ and $v = v_1 + v_2 + v_3$

$$\sigma_{xx} = \frac{A_1}{c_1} p(\lambda + 2\mu \cos^2 \alpha_1)\cos(pt + g_1 y) +$$

$$+ \frac{A_2}{c_1} p(\lambda + 2\mu \cos^2 \alpha_2)\cos(pt + g_2 y + \delta_1) -$$

$$- \frac{A_3}{c_2} p(\mu \sin 2\beta_2)\cos(pt + g_3 y + \delta_2).$$

When $g_1 = g_2 = g_3$ and $\delta_1 = \delta_2 = 0$ as before, this will be zero for all values of y and t, if the condition

$$(A_1 + A_2)(\lambda + 2\mu \cos^2\alpha_1) - A_3 \frac{c_1}{c_2} \mu \sin 2\beta_2 = 0$$

holds for the amplitudes. Substituting for λ and μ from the relation $(\lambda + 2\mu)/\mu = c_1^2/c_2^2 = \sin^2 \alpha_1/\sin^2 \beta_2$ this relation may be written

$$(A_1 + A_2)\cos 2\beta_2 \sin \alpha_1 - A_3 \sin \beta_2 \sin 2\beta_2 = 0. \qquad (2.42)$$

From equations (2.41) and (2.42) we may obtain the amplitudes of the two reflected waves, and since these equations apply to harmonic waves of any frequency, they will also be true for waves of arbitrary shape. At normal incidence $A_3 = 0$, and there is no reflected distortion wave. The amplitude of the reflected dilatation wave is then equal to that of the incident wave with a phase change of π on reflection at the boundary.

In Fig. 6 the values of the ratios A_2/A_1 and A_3/A_1 are plotted against the angle of incidence for a material for which Poisson's ratio is equal to $\frac{1}{3}$. c_1/c_2 is given by $[(\lambda + 2\mu)/\mu]^{\frac{1}{2}}$ and it may be seen from equation (2.5) that at this value of ν, $c_1/c_2 = 2$. The graphs show that the amplitude of the reflected distortion wave is a maximum at an angle of incidence of about $48°$ and its amplitude is then greater than that of the incident wave. The amplitude of the reflected dilatation wave is a minimum at an angle of incidence of about $65°$; at grazing incidence no distortion wave is reflected and A_2/A_1 again becomes unity. It should be noted that the energy flux of a distortion wave is less than that of a dilatation wave of the same displacement amplitude (the ratio between the energy fluxes is c_2/c_1) and also that since the distortion wave is reflected at a smaller angle than the angle of incidence, the width of a reflected beam of distortional waves will be greater than the width of the incident beam of dilatation waves and hence the energy density will be lower. When these

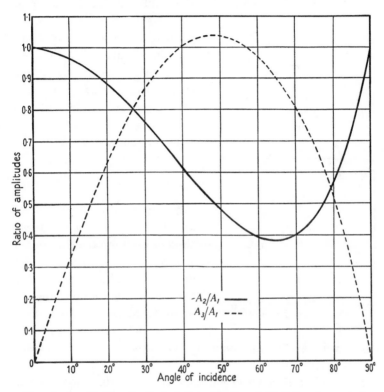

Fig. 6. Amplitudes of reflected distortional and dilatational waves at different angles of incidence for $v = \frac{1}{3}$.

two factors are taken into account, the curves in Fig. 6 show that the sum of the energies of the two reflected waves is equal to the energy of the incident dilatation wave.

We shall next consider the reflection of a wave of distortion incident on a free boundary. As before, we have a plane wave travelling parallel to the xy plane and impinging on a plane boundary which is in the yz plane. We take the angle of incidence to be β_1' (see Fig. 7). In order to treat this problem it is necessary to specify the direction of vibration of the wave. The displacements resulting from any wave of distortion may be considered as having been produced by the superposition of two component waves with their vibration directions at right angles. It will thus be sufficient to determine the conditions for the reflection of a wave with its vibrations parallel to the z-axis and one with its vibration direction perpendicular to this. The conditions for any other direction of vibration may then be found by combining the results.

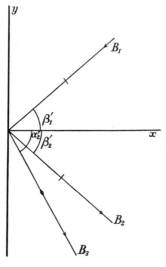

FIG. 7. Reflection of distortion wave at free boundary.

The boundary conditions to be satisfied are:

$$\sigma_{xx} = \sigma_{yx} = \sigma_{zx} = 0 \quad \text{at } x = 0.$$

For a wave with its vibration direction parallel to the z-axis there is no motion either in the x or y directions so that $u = 0$ and $v = 0$. Thus a wave of distortion of equal amplitude and opposite phase reflected at an angle equal to the angle of incidence will satisfy the boundary conditions and no dilatation wave is generated.

For a distortion wave with its vibration direction perpendicular to the z-axis, the treatment is analogous to that already described for an incident dilatation wave. There is here no motion in the z-direction, and the relevant conditions are $\sigma_{xx} = 0$ and $\sigma_{yx} = 0$ at the boundary. It is found that these conditions can only be satisfied by assuming that a dilatation wave as well as a distortion wave is reflected. The wave of distortion is reflected at an angle equal to the angle of incidence, whilst the dilatation wave is reflected at an angle α_2' where $\sin \alpha_2' / \sin \beta_1' = c_1/c_2$. If the amplitude of the incident distortion wave is B_1, of the reflected distortion wave B_2, and of the reflected dilatation wave B_3, the condition that $\sigma_{xx} = 0$ at $x = 0$ leads to:

$$(B_1 + B_2)\sin 2\beta_1' \sin \beta_1' - B_3 \sin \alpha_2' \cos 2\beta_1' = 0, \qquad (2.43)$$

and the condition $\sigma_{yx} = 0$ at $x = 0$ gives:

$$(B_1 - B_2)\cos 2\beta_1' - 2B_3 \sin \beta_1' \cos \alpha_2' = 0. \qquad (2.44)$$

From these two equations B_2/B_1 and B_3/B_1 can be found for any angle of incidence, and it may be seen that at normal incidence ($\beta_1' = 0$), $B_3 = 0$, and no dilatation wave is reflected.

Reflection and refraction at an interface between two media

As mentioned earlier, when an elastic wave of either type reaches a slip-free boundary, in general, four waves are generated. Two of these are refracted into the second medium, and two are reflected back. The treatment of this problem is similar to that already described for reflection at a free boundary, and therefore will not be worked out in detail here. Full accounts of the general case of reflection and refraction at a plane boundary between two media will be found in Knott (1899), Zoeppritz (1919), and Macelwane and Sohon (1936).

At the interface there are now four separate boundary conditions. These are that on both sides of the boundary the following four quantities must be equal:

  (i) the normal displacement;

  (ii) the tangential displacements;

  (iii) the normal stress;

  (iv) the tangential stress.

There are in this case five contributions to each displacement (one from the incident wave, two from the reflected waves and two from the refracted waves). If we consider a wave travelling in the xy plane and take the yz plane to be the interface, the four boundary conditions may be written:

  (i) $\sum u_a = \sum u_b,$

  (ii) $\sum v_a = \sum v_b$ and $\sum w_a = \sum w_b,$

  (iii) $\sum (\sigma_{xx})_a = \sum (\sigma_{xx})_b,$

or, substituting in terms of the displacements from equation (2.3):

$$\sum \left(\lambda\Delta + 2\mu\frac{\partial u}{\partial x}\right)_a = \sum \left(\lambda\Delta + 2\mu\frac{\partial u}{\partial x}\right)_b,$$

(iv) $\sum (\sigma_{yx})_a = \sum (\sigma_{yx})_b$ and $\sum (\sigma_{zx})_a = \sum (\sigma_{zx})_b,$

or, substituting from equation (2.3),

$$\sum \left[\mu\left(\frac{\partial v}{\partial x} + \frac{\partial u}{\partial y}\right)\right]_a = \sum \left[\mu\left(\frac{\partial v}{\partial x} + \frac{\partial u}{\partial y}\right)\right]_b$$

and

$$\sum \left[\mu\left(\frac{\partial w}{\partial x} + \frac{\partial u}{\partial z}\right)\right]_a = \sum \left[\mu\left(\frac{\partial w}{\partial x} + \frac{\partial u}{\partial z}\right)\right]_b.$$

The components of stress and strain in the first medium are here denoted by the suffix a, and those in the second medium are denoted by the suffix b. The above boundary conditions apply to the plane $x = 0$.

Consider a wave of dilatation travelling parallel to the xy plane and incident on the boundary at an angle α_1, and let the angles at which the dilatation waves are reflected and refracted be α_2 and α_3 respectively, whilst distortion waves are reflected and refracted at angles β_2 and β_3 respectively (see Fig. 8). It is found (Macelwane and Sohon, 1936) that the boundary conditions are satisfied if it is assumed that Huygens's principle can be applied to these waves: in other words, the wave front at any instant is on the envelope of a series of spherical wavelets sent out from points on the wave front at a previous instant. This construction leads, as in the case of light, to the relation:

$$\frac{\sin \alpha_1}{c_1} = \frac{\sin \alpha_2}{c_1} = \frac{\sin \beta_2}{c_2} = \frac{\sin \alpha_3}{c_3} = \frac{\sin \beta_3}{c_4}, \qquad (2.45)$$

c_1 and c_2 being the velocities of propagation of dilatation and distortion waves in the first medium, whilst c_3 and c_4 are the corresponding velocities in the second medium. Let the amplitude of the incident dilatation wave be A_1, those of the reflected and refracted dilatation waves be A_2 and A_4 respectively, and those of the corresponding distortion waves be A_3 and A_5 respectively (see Fig. 8).

Substituting in the boundary conditions leads to four relations between the amplitudes. Thus, condition (i) gives:

$$(A_1-A_2)\cos\alpha_1+A_3\sin\beta_2-A_4\cos\alpha_3-A_5\sin\beta_3 = 0; \quad (2.46)$$

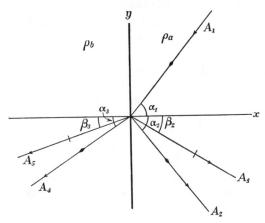

FIG. 8. Reflection and refraction of incident dilatation wave at plane interface.

the first part of condition (ii) (i.e. $\sum v_a = \sum v_b$) gives:

$$(A_1+A_2)\sin\alpha_1+A_3\cos\beta_2-A_4\sin\alpha_3+A_5\cos\beta_3 = 0; \quad (2.47)$$

condition (iii) becomes, since the motion is in the xy plane and $w = 0$,

$$\sum\left[(\lambda+2\mu)\,\frac{\partial u}{\partial x}+\lambda\,\frac{\partial v}{\partial y}\right]_a = \sum\left[(\lambda+2\mu)\,\frac{\partial u}{\partial x}+\lambda\,\frac{\partial v}{\partial y}\right]_b.$$

This leads to:

$$(A_1+A_2)c_1\cos 2\beta_2-A_3 c_2\sin 2\beta_2-A_4 c_3(\rho_b/\rho_a)\cos 2\beta_3-$$
$$-A_5 c_4(\rho_b/\rho_a)\sin 2\beta_3 = 0 \quad (2.48)$$

where ρ_a and ρ_b are the densities of the two media. (λ and μ have here been expressed in terms of the velocities of propagation and the densities of the media.)

The relevant part of condition (iv), namely

$$\sum (\sigma_{yx})_a = \sum (\sigma_{yx})_b,$$

leads to:

$$\rho_a c_2^2[(A_1-A_2)\sin 2\alpha_1-A_3(c_1/c_2)\cos 2\beta_2]-$$
$$-\rho_b c_4^2[A_4(c_1/c_3)\sin 2\alpha_3-A_5(c_1/c_4)\cos 2\beta_3] = 0. \quad (2.49)$$

By substituting from (2.45) the four simultaneous equations (2.46), (2.47), (2.48), and (2.49) may be solved to give the amplitudes of the reflected and refracted waves in terms of the amplitude of the incident dilatation wave.

For the case of normal incidence, α_1 is zero, and, from (2.45), all the other angles are zero; substituting in the four equations it may be seen that A_3 and A_5 vanish, so that only dilatation waves are generated. The solutions for A_2 and A_4 are then given by:

$$A_2 = A_1(\rho_b c_3 - \rho_a c_1)/(\rho_b c_3 + \rho_a c_1) \tag{2.50}$$

and

$$A_4 = A_1 2\rho_a c_1/(\rho_b c_3 + \rho_a c_1). \tag{2.51}$$

The amplitude of the reflected stress wave thus depends on the quantity $(\rho_b c_3 - \rho_a c_1)$ and no wave will be reflected at normal incidence when the product of the density and velocity is the same for the two media. This product ρc is sometimes known as the *characteristic impedance* of the medium. It may be seen from (2.50) that, when the characteristic impedance of the second medium is higher than that of the first, the amplitude of the displacement on reflection is of the same sign as that of the incident wave. Since, however, the direction of propagation is reversed on reflection, this corresponds to a change in phase of π in the vibrations. When the characteristic impedance of the second medium is lower than that of the first, the amplitude also changes in sign, so that there is no change in phase on reflection.

We now come to the reflection and refraction of distortion waves incident on a plane interface. We assume that the incident wave is of amplitude B_1, that it is travelling parallel to the xy plane, and that it is incident at an angle β_1' on a plane interface in the yz plane. As in the case of reflection at a free boundary, the direction of vibration of the incident distortion wave must be specified, and we must consider two separate cases, namely, where the vibrations are parallel to the z-axis, and where they are perpendicular to it and take place in the xy plane. When the vibration direction is parallel to the z-axis, there is no motion normal to the interface, and no dilatation waves are reflected or refracted. Let the amplitudes of the reflected and refracted distortion waves be B_2 and B_5; the former will be reflected at an angle equal to the angle of incidence, whilst the latter will be refracted at an angle β_3' where $\sin\beta_3'/\sin\beta_1' = c_4/c_2$. The only relevant boundary conditions are the second parts of (ii) and (iv), namely:

$$\sum w_a = \sum w_b, \quad \text{and} \quad \sum (\sigma_{zx})_a = \sum (\sigma_{zx})_b.$$

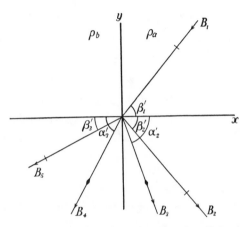

FIG. 9. Reflection and refraction of incident distortion
wave at plane interface.

These lead to the relations:

$$B_1 + B_2 - B_5 = 0 \qquad (2.52)$$

and
$$\rho_a \sin 2\beta_1'(B_1 - B_2) - B_5\rho_b \sin 2\beta_3' = 0. \qquad (2.53)$$

Where the vibrations of the incident wave of distortion are in the xy plane, four waves are generated (see Fig. 9). To satisfy the boundary conditions, the angles must here again obey the sine law as in equation (2.45), and, if the angles that the reflected and refracted dilatation waves make with the normal are α_2' and α_3' respectively, and their amplitudes are B_3 and B_4, we have:

$$\frac{\sin\beta_1'}{c_2} = \frac{\sin\beta_2'}{c_2} = \frac{\sin\alpha_2'}{c_1} = \frac{\sin\alpha_3'}{c_3} = \frac{\sin\beta_3'}{c_4}. \qquad (2.54)$$

The four boundary conditions in this case lead respectively to the following relations between the amplitudes:

$$(B_1-B_2)\sin\beta_1'+B_3\cos\alpha_2'+B_4\cos\alpha_3'-B_5\sin\beta_3' = 0, \quad (2.55)$$

$$(B_1+B_2)\cos\beta_1'+B_3\sin\alpha_2'-B_4\sin\alpha_3'-B_5\cos\beta_3' = 0, \quad (2.56)$$

$$c_2(B_1+B_2)\sin 2\beta_1'-B_3 c_1\cos 2\beta_1'+$$
$$+B_4 c_3(\rho_b/\rho_a)\cos 2\beta_3'-B_5 c_4(\rho_b/\rho_a)\sin 2\beta_3' = 0, \quad (2.57)$$

$$\rho_a c_2[(B_1-B_2)\cos 2\beta_1'-B_3(c_2/c_1)\sin 2\alpha_2']-$$
$$-\rho_b c_4[c_4/c_3 B_4\sin 2\alpha_3'+B_5\cos 2\beta_3'] = 0. \quad (2.58)$$

For normal incidence it may be seen that no dilatation waves are generated, and the relations simplify to:

$$B_1+B_2-B_5 = 0$$

and $\qquad\qquad \rho_a c_2(B_1-B_2)-\rho_b c_4 B_5 = 0.$

When the product of the velocity of distortion waves and the density is the same in both media at normal incidence $B_2 = 0$ and no wave of distortion is reflected.

The relations between the amplitudes for reflection at a free boundary may be derived from the general equations at an interface by putting ρ_b equal to zero. The conditions for equality of the displacements, (i) and (ii), cannot now be applied, but the equations obtained from conditions (iii) and (iv), i.e. (2.48), (2.49), (2.57), and (2.58), may be seen to simplify to equations (2.41) to (2.44) obtained for a free boundary in the previous section.

Total reflection

In the previous two sections the boundary conditions have been satisfied by using relations such as (2.45) between the angles of incidence, reflection, and refraction. When, however, the velocity of propagation for the reflected or refracted wave is higher than that of the incident wave, there will be a critical angle of incidence which makes the angle of reflection or refraction $\frac{1}{2}\pi$. For angles of incidence greater than this, the relation breaks down, and the situation is similar to that known as total internal reflection in optics.

In the case of a free boundary, this will only occur for a wave of distortion incident at an angle whose sine is greater than

$\sqrt{[\mu/(\lambda+2\mu)]}$. For a wave impinging on an interface between two media, there will also be critical angles of incidence for the refracted waves; thus, when c_3 is greater than c_1, an incident dilatation wave will generate a dilatation wave in the second medium only when the sine of the angle of incidence is less than c_1/c_3. For angles of incidence greater than the critical angle, the problem must be treated, as in the case of optics, in terms of complex quantities. It is found that, in place of a reflected or refracted plane wave, a disturbance is now set up which decays exponentially with distance from the interface. This wave does not carry away energy from the boundary, and the energy of the incident waves is divided between the remaining reflected and refracted waves. The presence of this damped wave does, however, result in a change in phase in the other waves generated.

To illustrate the treatment of total reflection, we shall consider again the simple case of a plane harmonic distortion wave incident on a free boundary (see Fig. 7). As before, let the wave travel parallel to the xy plane, with the vibrations taking place in this plane. Let the angle of incidence be β_1', and let a dilatation wave be reflected at an angle α_2', where

$$\sin\alpha_2'/\sin\beta_1' = c_1/c_2 = [(\lambda+2\mu)/\mu]^{\frac{1}{2}}.$$

If we call the displacement produced by the incident wave parallel to its wave front Ψ_1'', the corresponding displacement for the reflected distortion wave Ψ_2'', and that for the reflected dilatation wave Φ_2', we may write:

$$\Psi_1'' = \text{real part of } B_1 \exp[i(pt+f_1'x+g_1'y)],$$

where $\quad f_1' = p(\cos\beta_1')/c_2 \quad$ and $\quad g_1' = p(\sin\beta_1')/c_2$

$$\Psi_2'' = \text{real part of } B_2\exp[i(pt-f_1'x+g_1'y+\delta_1)],$$

and $\quad \Phi_2' = \text{real part of } B_3\exp[i(pt-f_2'x+g_2'y+\delta_2)]$

where $\quad f_2' = p(\cos\alpha_2')/c_1 \quad$ and $\quad g_2' = p(\sin\alpha_2')/c_1$.

Now, if $\sin\beta_1' > c_2/c_1$, $\sin\alpha_2'$ becomes greater than unity; hence $\cos\alpha_2'$ is a pure imaginary, and f_2' may be equated to $-im$, where $m = (p/c_1)[(c_1/c_2)^2\sin^2\beta_1'-1]^{\frac{1}{2}}$. Thus Φ_2' becomes $B_3 e^{-mx}\cos(pt+g_2'y)$. This corresponds to a wave, the amplitude of which decays exponentially with distance from the interface and whose phase is independent of x. The amplitude of the reflected distortion wave is now equal to that of the incident wave, i.e. $B_1 = B_2$ and, to satisfy the boundary conditions, δ_1 can no longer be taken as 0 or π, but is a rather involved function of β_1 and (c_1/c_2).

As mentioned earlier, the relations obtained between the amplitudes and directions and phases of incident, reflected, and refracted sinusoidal waves are independent of the wavelength, and the results will therefore apply whatever the shapes of the waves. When total reflection takes place, however, this no longer holds and the analysis becomes very involved. Friedlander (1948) discusses the problem of a plane distortion wave of arbitrary shape incident on an interface when the angle of incidence exceeds the critical angle.

Propagation in an aeolotropic medium

In deriving the equations of motion for a solid medium (2.7), it was pointed out that these equations would apply whatever the stress-strain relations of the medium might be. The wave equations for an isotropic elastic solid were then obtained by substitution from the appropriate elastic equations (2.3). In order to investigate the propagation of elastic waves in an aeolotropic solid, e.g. a single crystal, the general stress-strain relations (2.2) must be used. These give the stress components in terms of the strain components, and the strains in turn may be expressed in terms of the displacements, u, v, and w, by substitution from equation (2.1).

Consider a plane wave propagated with velocity c through an aeolotropic elastic medium, the normal to the wave having direction cosines l, m, and n. The displacements u, v, and w, will then be functions of a single parameter ψ, where

$$\psi = lx + my + nz - ct,$$

so that
$$\epsilon_{xx} = \frac{\partial u}{\partial x} = l\frac{\partial u}{\partial \psi},$$

$$\epsilon_{yy} = m\frac{\partial v}{\partial \psi}, \qquad \epsilon_{zz} = n\frac{\partial w}{\partial \psi}, \qquad \epsilon_{yz} = m\frac{\partial w}{\partial \psi} + n\frac{\partial v}{\partial \psi},$$

$$\epsilon_{zx} = n\frac{\partial u}{\partial \psi} + l\frac{\partial w}{\partial \psi} \quad \text{and} \quad \epsilon_{xy} = l\frac{\partial v}{\partial \psi} + m\frac{\partial u}{\partial \psi}.$$

Thus the first equation in (2.2) becomes:

$$\sigma_{xx} = (c_{11}l + c_{16}m + c_{15}n)\frac{\partial u}{\partial \psi} + (c_{16}l + c_{12}m + c_{14}n)\frac{\partial v}{\partial \psi} +$$

$$+ (c_{15}l + c_{14}m + c_{13}n)\frac{\partial w}{\partial \psi}$$

and

$$\frac{\partial \sigma_{xx}}{\partial x} = l[(c_{11}l + c_{16}m + c_{15}n)u'' + (c_{16}l + c_{12}m + c_{14}n)v'' +$$

$$+ (c_{15}l + c_{14}m + c_{13}n)w''],$$

where u'', v'', w'' denote the second differential coefficients of u, v, and w with respect to ψ. Similar equations may be obtained for $\partial\sigma_{xy}/\partial y$ and $\partial\sigma_{xz}/\partial z$ and since $\partial^2 u/\partial t^2 = c^2 u''$, the first equation of motion (2.7) may be written:

$$\rho c^2 u'' = Au'' + Hv'' + Gw''$$

and the other two equations (2.7) become:

$$\rho c^2 v'' = Hu'' + Bv'' + Fw''$$

and

$$\rho c^2 w'' = Gu'' + Fv'' + Cw''$$

(2.59)

where:

$$A = c_{11}l^2 + c_{66}m^2 + c_{55}n^2 + 2c_{16}lm + 2c_{56}mn + 2c_{15}nl,$$

$$B = c_{66}l^2 + c_{22}m^2 + c_{44}n^2 + 2c_{26}lm + 2c_{24}mn + 2c_{46}nl,$$

$$C = c_{55}l^2 + c_{44}m^2 + c_{33}n^2 + 2c_{45}lm + 2c_{34}mn + 2c_{35}nl,$$

$$F = c_{56}l^2 + c_{24}m^2 + c_{34}n^2 + (c_{25} + c_{46})lm +$$

$$+ (c_{23} + c_{44})mn + (c_{36} + c_{45})nl,$$

$$G = c_{15}l^2 + c_{46}m^2 + c_{35}n^2 + (c_{14} + c_{56})lm +$$

$$+ (c_{36} + c_{45})mn + (c_{13} + c_{55})nl,$$

$$H = c_{16}l^2 + c_{26}m^2 + c_{45}n^2 + (c_{12} + c_{66})lm +$$

$$+ (c_{25} + c_{46})mn + (c_{14} + c_{56})nl.$$

Eliminating u'', v'', and w'' from (2.59) gives the determinantal equation:

$$\begin{vmatrix} (A - \rho c^2) & H & G \\ H & (B - \rho c^2) & F \\ G & F & (C - \rho c^2) \end{vmatrix} = 0.$$

(2.60)

This is a cubic in c^2 which has three positive roots for any real elastic solid. In general, these roots will be different, and correspond to three distinct velocities of propagation. The values of these velocities will depend on the twenty-one elastic constants of the material, and on the direction of propagation as defined by l, m, and n. The wave surface will have three sheets as compared with the two sheets of the Fresnel surface for the propagation of light in a crystalline medium. It may be shown (Kelvin, 1904, p. 135), that, when the three velocities of propagation are different, equations (2.59) imply that the vibration directions accompanying the three velocities are mutually perpendicular. When two of the velocities of propagation are equal the vibrations corresponding to them will combine into a single wave motion, and this will take place in a plane perpendicular to the third vibration direction. When this occurs, the combined motion may be, as in the case of light, in the form of plane polarization, elliptic polarization, or circular polarization, depending on the phase relations between the two components of the vibration and on their amplitudes.

CHAPTER III

PROPAGATION IN BOUNDED ELASTIC MEDIA

In the last chapter the equations of motion for an isotropic solid medium were obtained in terms of the displacements (equations (2.8), (2.9), and (2.10)). In theory the propagation of stress waves in any bounded isotropic solid can be derived by solving these equations for the appropriate boundary conditions. From the treatment of the reflection of plane elastic waves at a plane interface, it will have been seen that, where there are several free surfaces, the problem is not likely to be an easy one, and, except for the very simplest cases, no exact solutions have in fact been obtained.

In this chapter the propagation of stress waves along a cylindrical bar will be considered first, as this is a problem which has been investigated most fully theoretically and on which there are also some experimental data. Before examining the problem in terms of the exact elastic equations, we shall consider the simple treatment which applies to the propagation of waves the lengths of which are large compared with the diameter of the bar.

There are three different types of vibration which occur in thin rods or bars; these are classed as longitudinal, torsional, and lateral. In longitudinal vibrations, elements of the rod extend and contract, but there is no lateral displacement of the axis of the rod. In torsional vibrations, each transverse section of the rod remains in its own plane and rotates about its centre, the axis of the rod remaining undisturbed. Lastly, lateral vibrations correspond to the flexure of portions of the rod, elements of the central axis moving laterally during the motion.

Longitudinal vibrations of rods

If we consider each plane cross-section of the rod to remain plane during the motion and the stress over it to be uniform, we may obtain the equation of motion directly.

For consider a small element PQ of length δx and let the cross-sectional area of the rod be A (see Fig. 10). If the stress on the face passing through P is σ_{xx} the stress on the other face will be given by $\sigma_{xx}+(\partial\sigma_{xx}/\partial x)\,\delta x$, and if the displacement of the element is given by u, we have from Newton's second law of motion:

$$\rho A\,\delta x\,\frac{\partial^2 u}{\partial t^2}=A\,\frac{\partial\sigma_{xx}}{\partial x}\,\delta x \tag{3.1}$$

(where ρ is the density of the rod).

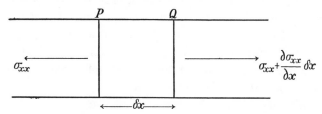

FIG. 10. Forces acting on element of bar in longitudinal motion.

Now the ratio between the stress σ_{xx} and the strain $\partial u/\partial x$ in the element is Young's modulus E, so that (3.1) may be written:

$$\rho\,\frac{\partial^2 u}{\partial t^2}=E\,\frac{\partial^2 u}{\partial x^2}. \tag{3.2}$$

This equation is of the same form as (2.15), and corresponds to the propagation of longitudinal waves along the bar with velocity $\sqrt{(E/\rho)}$.

The solution of (3.2) may be written as:

$$u=f(c_0 t-x)+F(c_0 t+x), \tag{3.3}$$

where $$c_0=\sqrt{(E/\rho)} \tag{3.4}$$

and F and f are here arbitrary functions depending on the initial conditions. The function f corresponds to a wave travelling in the direction of increasing x, whilst the function F corresponds to a wave travelling in the opposite direction. In deriving (3.2), it has not been assumed that the rod is necessarily cylindrical, and the equation will apply equally for thin bars or rods of any uniform cross-section.

The reason that the above treatment is only approximate is that, in deriving it, it has been assumed that plane transverse sections of the rod remain plane during the passage of the stress waves, and the stress acts uniformly over each section. The longitudinal expansions and contractions of sections of the rod will, however, necessarily result in lateral contractions and expansions, the ratio between these lateral and longitudinal strains being given by Poisson's ratio ν. This lateral motion will result in a non-uniform distribution of stress across the sections of the bar, and plane transverse sections will become distorted. The effect of this lateral motion in cylindrical bars is discussed later, and it will be shown that it becomes of importance when the operative wavelengths are of the same order as the diameter of the bar.

We shall, however, consider first the consequences of equation (3.3), which may be applied to the propagation of elastic waves the lengths of which are large compared with the lateral dimensions of the bar. For simplicity we take a wave travelling in the direction of decreasing x, we then have:

$$u = f(c_0 t + x). \tag{3.5}$$

Differentiating both sides of (3.5) with respect to x we have:

$$\frac{\partial u}{\partial x} = f'(c_0 t + x), \tag{3.6}$$

where f' denotes differentiation with respect to the argument $(c_0 t + x)$.

Similarly, if we differentiate (3.5) with respect to t we have:

$$\frac{\partial u}{\partial t} = c_0 f'(c_0 t + x). \tag{3.7}$$

Thus, from (3.6) and (3.7),

$$\frac{\partial u}{\partial t} = c_0 \frac{\partial u}{\partial x}.$$

Now $\partial u / \partial x$ is equal to σ_{xx}/E, so that

$$\sigma_{xx} = \left(\frac{E}{c_0}\right) \frac{\partial u}{\partial t} = \rho c_0 \frac{\partial u}{\partial t}. \tag{3.8}$$

Thus equation (3.8) shows that there is a linear relation between the stress at any point and the particle velocity, the ratio between them, ρc_0, corresponding to the *characteristic impedance* as defined in connexion with equation (2.50) and (2.51). By analogy with the case of electrical conduction, equation (3.8) is sometimes referred to as the mechanical counterpart of Ohm's law, and the value of the characteristic impedance of the material in c.g.s. units is sometimes given in 'acoustical ohms'.

Since the velocity of propagation c_0 is independent of the frequency of the stress waves, a stress pulse whose operative Fourier components have wavelengths long compared with the diameter of the bar will travel down it without dispersion. When such a pulse reaches a free end of the bar, it will be reflected, and to find the nature of the reflected pulse we apply the boundary condition that there is no stress normal to the end face of the bar. If we take the displacement due to the incident pulse to be given by

$$u_1 = F(c_0 t + x), \tag{3.9}$$

and that due to the reflected pulse to be

$$u_2 = f(c_0 t - x), \tag{3.10}$$

the stresses produced by the two pulses will be

$$E\frac{\partial u_1}{\partial x} \quad \text{and} \quad E\frac{\partial u_2}{\partial x},$$

so that the resultant stress will be:

$$E\left(\frac{\partial u_1}{\partial x} + \frac{\partial u_2}{\partial x}\right) = E[F'(c_0 t + x) - f'(c_0 t - x)]. \tag{3.11}$$

If we measure x from the end of the bar the condition that the end is free from stress is

$$F'(c_0 t) - f'(c_0 t) = 0. \tag{3.12}$$

Thus the shape of the reflected pressure pulse is the same as that of the incident pulse, but is opposite in sign; thus a compression pulse will be reflected as a similar pulse of tension. The displacement of any point on the bar is given by $u_1 + u_2$, and at the free end of the bar where $x = 0$ this will be $2F(c_0 t)$, so that the displacements, and hence the particle velocities at the end of the bar, will be twice their corresponding values when the pulse is travelling along the bar.

When a pressure pulse is reflected at a fixed boundary at the end of a bar, the boundary condition is that the displacement is zero at $x = 0$. From (3.9) and (3.10) the total displacement is given by:

$$u_1 + u_2 = F(c_0 t + x) + f(c_0 t - x) \qquad (3.13)$$

and this is zero at $x = 0$, so that $f(c_0 t)$ is equal to $-F(c_0 t)$; u_2, the displacement in the reflected pulse is thus equal and opposite to the displacement in the incident pulse. The stress set up by the reflected pulse is given by $E\, \partial u_2/\partial x$ and this is now found to be equal to $E\, \partial u_1/\partial x$, the stress for the incident pulse. A pressure pulse is thus reflected at a fixed boundary unchanged since the pressure distribution remains the same, and both the direction of the displacement and the direction of propagation are reversed. The stresses produced by the incident and reflected pulses thus add up on a fixed boundary, and the values of the resultant stress are double the corresponding values when the pulse is travelling along the bar.

As an illustration of the results of the elementary treatment we shall now consider the motion of a freely suspended bar AC at the back end C of which a constant force P has been applied for a short interval of time δt. The total momentum imparted to the bar will be $P\,\delta t$, and if the mass of the bar is M the centre of gravity will then move with constant velocity V where $MV = P\,\delta t$. Now at the instant when the force is removed, a length $c_0\,\delta t$ of the bar will be compressed, and the remainder will be at rest and unstrained. The stress pulse will remain of constant length and will travel along the bar with velocity c_0. (It may be seen from equation (3.8) that the particle velocity produced by the pulse is $P/(\rho c_0 A)$ and the mass travelling with this velocity is $\rho A c_0\,\delta t$, so that the momentum is $P\,\delta t$ as before.) When the pulse reaches the front face A of the bar, it is reflected as a pulse of tension which travels back with velocity c_0. This tension pulse is reflected at the back face of the bar as a compression pulse and the whole cycle is then repeated.

The pulse will travel the length of the bar L in time L/c_0 and Fig. 11 illustrates the motion when $\delta t < 2L/c_0$. A and C are the displacement-time curves of the front and back faces

which move in jerks at intervals of $2L/c_0$. B is the curve for
the midpoint of the bar which is set into motion twice as often,
since the pulse passes through it twice for each reflection at an
end face. D, the curve for the centre of gravity of the bar, is a

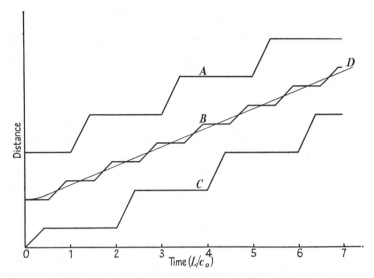

FIG. 11. Displacement-time curve of bar showing discontinuous motion produced by repeated reflections of longitudinal pressure pulse. Motion of (A)
front face of bar, (B) mid-point of bar, (C) back face of bar, (D) centre of
gravity of bar. (For clarity the magnitudes of the displacements in the figure
have been vastly exaggerated relative to the length of the bar.)

parabola for time δt, whilst the constant force P is being applied,
and after that is a straight line.

The necessarily constant velocity of the centre of gravity,
once the force has been removed, is associated with the difference
between the mass per unit length in the undisturbed region of
the bar and in the length $c_0\,\delta t$ containing the pulse. Thus, either
a compression pulse moving forwards, or a tension pulse moving
backwards results in a forward movement of the centre of
gravity. From curves B and D it may be seen that for each cycle
of duration $2L/c_0$ the mid-point of the bar and its centre of
gravity coincide four times. Two of these occasions are when
the pulse is passing through the centre of the bar so that the

region of abnormal linear density is evenly distributed about the centre. The other two are when just half the pulse has been reflected at one of the end faces. The stresses produced by the incident and reflected halves of the pulse then just cancel out and the density is uniform over the whole bar.

In practice the length of the pulse will continually increase as it traverses the bar. This will occur because of the dispersion produced by the radial motion of the bar, and also as a result of internal friction. Both these effects are discussed more fully later. The steps shown in Fig. 11 will consequently become rounded and eventually the length of the pulse will exceed that of the bar. The problem can then best be considered as the whole bar moving with constant velocity and at the same time performing free longitudinal oscillations about its centre of gravity. These oscillations slowly decay, leaving only the forward movement which is the motion considered in rigid dynamics.

Torsional vibrations of rods

In the torsional vibrations of a cylindrical bar, each transverse section remains in its own plane and rotates about its centre. The equation of motion may be obtained by considering the forces acting on an infinitesimally short element of the bar PQ of length δx (see Fig. 12).

Let the twisting couple acting on the section through P be C, and the opposing couple acting on the section through Q is then $C + (\partial C/\partial x)\,\delta x$. If the mean angle through which the element rotates about its centre is θ, we may equate the resultant couple to the product of the moment of inertia of the element I' and its angular acceleration. We thus have:

$$\frac{\partial C}{\partial x}\,\delta x = I'\,\frac{\partial^2 \theta}{\partial t^2}. \tag{3.14}$$

Now, if two opposing couples, each of magnitude C, act at opposite ends of a cylinder of length x', and radius r, they produce a relative angular rotation between the two end faces of θ', where

$$C = \tfrac{1}{2}\pi\mu r^4 \theta'/x'. \tag{3.15}$$

(This relation can be obtained by considering the cylinder as being divided into a series of infinitely thin coaxial tubes and integrating.) Hence, if the relative rotation between the sections through P and Q is $\delta\theta$, we find from (3.15) that in the limit:

$$C = \tfrac{1}{2}\pi\mu r^4 \frac{\partial\theta}{\partial x}.\qquad(3.16)$$

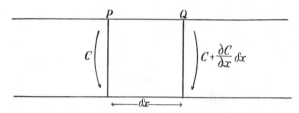

FIG. 12. Couples acting on element of bar in torsional motion.

The moment of inertia I' of the cylindrical element PQ about its axis is given by:

$$I' = \tfrac{1}{2}\pi\rho r^4\,\delta x.\qquad(3.17)$$

Substituting for C and I' from (3.16) and (3.17), in (3.14) we obtain:

$$\mu\,\frac{\partial^2\theta}{\partial x^2} = \rho\,\frac{\partial^2\theta}{\partial t^2}.\qquad(3.18)$$

This is once again the wave equation (2.19), and shows that torsional waves are propagated down a cylindrical bar with a velocity $\sqrt{(\mu/\rho)}$, which is identical with the velocity of propagation of waves of distortion in an infinite medium as found in the previous chapter.

It will be shown later that equation (3.18) may be derived from the general elastic relations, and, unlike equation (3.2) for longitudinal waves, (3.18) is an exact representation of the propagation of torsional vibrations along a circular cylinder when each section of the cylinder rotates as a whole. A torsional pulse of vibrations in this mode is thus transmitted along a cylindrical bar without dispersion so long as the material of the bar is perfectly elastic.

Flexural vibrations of rods

The theory of the flexural vibration of bars is more difficult than that of the two types of vibration already considered, since the elastic deformations involved are more complex, and even the elementary theory shows that the velocity of flexural waves depends on their wavelength. Discussions of the exact solution of the problem will be found in Timoshenko (1921), Prescott (1942), Hudson (1943), Cooper (1947), and Davies (1948).

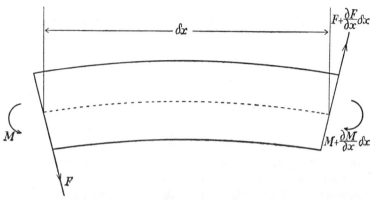

FIG. 13. Forces and couples acting on element of bar in flexural vibration.

As a result of the dispersion associated with this type of wave motion, the propagation of pulses of flexural waves has received little attention experimentally, and the theory has mainly been used to determine the period of free lateral vibration of bars. The resonant frequency of a vibrating cantilever gives an accurate method of measuring Young's modulus for a material dynamically, and has been used in the study of the elastic properties of solids (see, for example, Davies and James (1934), Grime and Eaton (1937), and Nolle (1948)).

In the simplest theory of flexural vibrations of bars of arbitrary but uniform cross-section it is assumed that the motion of each element of the bar is purely one of translation in a direction perpendicular to the axis of the bar.

Fig. 13 shows the forces acting on a very small element of the bar PQ of length δx, which has been bent in the xz plane. The bending moment varies along the bar, and is taken to have value M at P, whilst its value at Q may be written

$$M + \frac{\partial M}{\partial x} \, \delta x.$$

The bending moment will be balanced by shearing forces acting parallel to the z-axis. The shearing force on the section through P is taken to be F, whilst that through Q is

$$F + \frac{\partial F}{\partial x} \, \delta x.$$

The equation of motion of the element in the z-direction is then:

$$\rho A \, \delta x \, \frac{\partial^2 w}{\partial t^2} = \frac{\partial F}{\partial x} \, \delta x$$

or

$$\rho A \, \frac{\partial^2 w}{\partial t^2} = \frac{\partial F}{\partial x} \tag{3.19}$$

(where ρ is the density of the bar, A its cross-sectional area, and w the displacement in the z-direction).

To solve (3.19) we must use the elastic relations to express F in terms of w and the elastic constants of the material. Taking moments round an axis in the y-direction through the centre of the element, we have:

$$\frac{\partial M}{\partial x} \delta x + \left(2F + \frac{\partial F}{\partial x} \delta x \right) \frac{\delta x}{2} = 0; \tag{3.20}$$

in the limit where δx is infinitesimal this becomes:

$$F = -\frac{\partial M}{\partial x}. \tag{3.21}$$

To obtain the relation between M and w, we consider the element of the bar to be composed of a collection of parallel filaments; those above the neutral surface are stretched, whilst those below it are compressed. It is then found that if the radius of curvature of the neutral surface is R and the second moment of the cross-section about a diameter in this surface is I:

$$M = \frac{EI}{R}. \tag{3.22}$$

For small deformations $1/R$ will be given by $\partial^2 w / \partial x^2$, so that (3.21) becomes:

$$F = -EI \, \frac{\partial^3 w}{\partial x^3} \tag{3.23}$$

and substituting in (3.19) we have:

$$\rho A \frac{\partial^2 w}{\partial t^2} = -EI \frac{\partial^4 w}{\partial x^4}.$$

This may be written:

$$\frac{\partial^2 w}{\partial t^2} = -c_0^2 K^2 \frac{\partial^4 w}{\partial x^4}, \tag{3.24}$$

where K is the radius of gyration of the cross-section about an axis in the neutral surface perpendicular to the axis of the bar so that $I = K^2 A$. For a cylindrical bar of radius a, $K = a/2$. Equation (3.24) is the wave equation for flexural vibrations, and it may be seen, on substitution, that a solution of the form $w = f(x-ct)$ or $w = f(x+ct)$ will not in general satisfy this equation. Thus a flexural disturbance of arbitrary shape will not be propagated along a bar without dispersion.

Now if sinusoidal flexural waves are propagated along the bar with velocity c' we have:

$$w = D \cos(pt - fx), \tag{3.25}$$

where D is the amplitude, $f = 2\pi/\Lambda$, and $p = 2\pi c'/\Lambda$, Λ being the wavelength. Differentiating (3.25) and substituting in (3.24), we obtain:

$$p^2 = c_0^2 K^2 f^4$$

or

$$c' = \frac{2\pi c_0 K}{\Lambda}. \tag{3.26}$$

Thus c' is inversely proportional to the wavelength, and waves of infinitely short wavelength would, according to (3.26), travel with infinite velocity.

Since (3.25) implies an infinite wave train it gives information only regarding the difference in phase between the vibrations along a bar when a sinusoidal wave is propagated, and c' is termed the wave velocity or *phase velocity*. To determine the rate at which the energy of a pulse of flexural vibrations is propagated we must find the *group velocity* c_g'. This is defined as the velocity with which a packet of waves is propagated, the wavelengths of the component waves of the packet being close to Λ. The relation between c'_g and c' is:

$$c'_g = c' - \Lambda \frac{dc'}{d\Lambda} \tag{3.27}$$

(see for example Stephens and Bate, 1950, p. 76). Thus from (3.26) we have:

$$c'_g = c' + \Lambda \frac{2\pi c_0 K}{\Lambda^2} = 2c'. \tag{3.28}$$

The group velocity for flexural waves is, according to (3.28) twice the phase velocity, and thus it too becomes infinite for a pulse composed of infinitely short waves. The result that a flexural pulse can be transmitted with infinite velocity is unsound physically, and in fact (3.26) and (3.28) apply only to waves for which Λ is large compared with K, the radius of gyration of the bar.

The reasons why the above treatment breaks down when the wavelength is comparable with the lateral dimensions of the bar are:

(i) the assumption that the motion is purely one of translation in the z-direction is unjustified for short wavelengths since the *rotary motion* of the sections of the bar must then also be taken into account; and

(ii) the assumption that longitudinal sections of elements of the bar remain rectangular in shape during the motion is also unjustified for vibrations whose wavelength is comparable with the thickness of the bar.

To allow for (i), a term must be added to equation (3.20) to allow for the *rotary inertia* of the element. Thus, in taking moments, we must equate the resultant couple to the product of the moment of inertia of the element and its angular acceleration. Equation (3.20) thus becomes:

$$\left(F + \frac{\partial M}{\partial x}\right) \delta x = \rho I \frac{\partial^2 \alpha}{\partial t^2} \delta x, \tag{3.29}$$

where I is the second moment of a cross-section about an axis as before, and α is the angle through which the section has rotated. For small deformations α will be given by $\partial w/\partial x$ so that (3.29) becomes

$$F = -\frac{\partial M}{\partial x} + \rho I \frac{\partial^3 w}{\partial x \partial t^2}$$

and

$$\frac{\partial F}{\partial x} = -\frac{\partial^2 M}{\partial x^2} + \rho I \frac{\partial^4 w}{\partial x^2 \partial t^2}. \tag{3.30}$$

Substituting for $\partial F/\partial x$ from (3.19) and for M from (3.22), we have:

$$\rho A \frac{\partial^2 w}{\partial t^2} = -EI \frac{\partial^4 w}{\partial x^4} + \rho I \frac{\partial^4 w}{\partial x^2 \partial t^2}$$

or

$$c_0^2 K^2 \frac{\partial^4 w}{\partial x^4} - K^2 \frac{\partial^4 w}{\partial x^2 \partial t^2} + \frac{\partial^2 w}{\partial t^2} = 0. \tag{3.31}$$

This equation is similar to (3.24) with an additional term to allow for rotary inertia, and if we try (3.25) as a solution we obtain:

$$c' = c_0 \left(1 + \frac{\Lambda^2}{4\pi^2 K^2}\right)^{-\frac{1}{2}} \tag{3.32}$$

and from (3.27):

$$c_g' = c_0 \left(1 + \frac{\Lambda^2}{4\pi^2 K^2}\right)^{-\frac{1}{2}} \left(1 + \frac{1}{1 + 4\pi^2 K^2/\Lambda^2}\right). \tag{3.33}$$

(3.32) and (3.33) give the same results as (3.26) and (3.28) when K/Λ is small, whilst both c' and c_g' approach c_0 when K/Λ is large.

The correction for rotary inertia (i) was applied by Rayleigh (1894) and the resultant equation (3.31) is physically much more satisfactory in that it does not lead to infinite values for the group velocity. Timoshenko (1921) has shown, however, that correction (ii) for the shearing of elements of the bar is as important as (i). This correction is due to the fact that the shear forces F, shown in Fig. 13, will distort each element, and as a result the slope of the axis of the bar in the displaced position is the sum of the angle through which the element has been rotated and the angle through which it has been sheared, and when this is taken into account the equation (3.31) becomes:

$$c_0^2 K^2 \frac{\partial^4 w}{\partial x^4} - K^2(1+\epsilon') \frac{\partial^4 w}{\partial x^2 \partial t^2} + \frac{\epsilon' K^2}{c_0^2} \frac{\partial^4 w}{\partial t^4} + \frac{\partial^2 w}{\partial t^2} = 0. \tag{3.34}$$

This is the same as (3.31) except for the two additional terms involving ϵ'. ϵ' is a non-dimensional quantity equal to $2(1+\nu)/R'$, R' being a constant depending on the shape of the cross-section of the bar, and ν Poisson's ratio for the material. For a bar of circular cross-section $R' = 10/9$. Prescott (1942) has derived (3.34) by a different method from that used by Timoshenko. It will be shown in the next section (see Figs. 16 and 17) that the numerical results for the velocity obtained from this equation for a cylindrical bar are in very good agreement with those found by Hudson (1943), who calculated these velocities from the general elastic equations.

The Pochhammer equations for cylindrical bars

As mentioned earlier, it is possible in theory to solve any problem of vibration or stress propagation in an elastic solid by inserting the appropriate boundary conditions in the equations of motion (2.8), (2.9), and (2.10) of the previous chapter. In practice, however, exact solutions have not been obtained for even the simple case of the vibrations of a cylinder of finite length, although in this particular instance solutions may be derived which give results very close to the truth, so long as the length of the cylinder is large compared with its diameter. The problem was first investigated in terms of the general elastic equations by Pochhammer (1876) and independently by Chree (1889). An account of Pochhammer's treatment will be found in Love (1927), p. 287.

It is extremely difficult to insert the boundary conditions for a circular cylinder in the equations of motion in Cartesian co-ordinates (2.8), (2.9), and (2.10), and it is therefore necessary to transform these equations into cylindrical polar coordinates. This is done in the Appendix, and it is shown there that if the cylindrical coordinates are taken as r, θ, and z, and the corresponding displacements are u_r, u_θ, and u_z the equations may be written:

$$\rho \frac{\partial^2 u_r}{\partial t^2} = (\lambda + 2\mu) \frac{\partial \Delta}{\partial r} - \frac{2\mu}{r} \frac{\partial \bar{\omega}_z}{\partial \theta} + 2\mu \frac{\partial \bar{\omega}_\theta}{\partial z}, \qquad (3.35)$$

$$\rho \frac{\partial^2 u_\theta}{\partial t^2} = (\lambda + 2\mu) \frac{1}{r} \frac{\partial \Delta}{\partial \theta} - 2\mu \frac{\partial \bar{\omega}_r}{\partial z} + 2\mu \frac{\partial \bar{\omega}_z}{\partial r}, \qquad (3.36)$$

$$\rho \frac{\partial^2 u_z}{\partial t^2} = (\lambda + 2\mu) \frac{\partial \Delta}{\partial z} - \frac{2\mu}{r} \frac{\partial}{\partial r} (r \bar{\omega}_\theta) + \frac{2\mu}{r} \frac{\partial \bar{\omega}_r}{\partial \theta}, \qquad (3.37)$$

where, as before, Δ is the dilatation which in cylindrical coordinates is given by:

$$\Delta = \frac{1}{r}\frac{\partial(ru_r)}{\partial r} + \frac{1}{r}\frac{\partial u_\theta}{\partial \theta} + \frac{\partial u_z}{\partial z} \qquad (3.38)$$

and $\bar{\omega}_r$, $\bar{\omega}_\theta$, and $\bar{\omega}_z$ are the components of the rotation about three orthogonal directions. These directions are taken along the radius vector r, perpendicular to the rz plane and parallel to the z-axis respectively. The three components are given by:

$$\left. \begin{aligned} 2\bar{\omega}_r &= \frac{1}{r}\frac{\partial u_z}{\partial \theta} - \frac{\partial u_\theta}{\partial z} \\[2mm] 2\bar{\omega}_\theta &= \frac{\partial u_r}{\partial z} - \frac{\partial u_z}{\partial r} \\[2mm] 2\bar{\omega}_z &= \frac{1}{r}\left[\frac{\partial(ru_\theta)}{\partial r} - \frac{\partial u_r}{\partial \theta}\right] \end{aligned} \right\} . \qquad (3.39)$$

The three equations (3.39) lead to the identical relation:

$$\frac{1}{r}\frac{\partial(r\bar{\omega}_r)}{\partial r} + \frac{1}{r}\frac{\partial\bar{\omega}_\theta}{\partial \theta} + \frac{\partial\bar{\omega}_z}{\partial z} = 0. \qquad (3.40)$$

In applying equations (3.35), (3.36), and (3.37) to the vibrations of a circular cylinder we shall take the z-axis to be along the axis of the cylinder. At the surface of the cylinder the three stress components acting radially will vanish; these, by analogy with the symbols used in the Cartesian notation, may be denoted by σ_{rr}, $\sigma_{r\theta}$, and σ_{rz}. The relations between these and the strains are given by:

$$\left. \begin{aligned} \sigma_{rr} &= \lambda\Delta + 2\mu\frac{\partial u_r}{\partial r} \\[2mm] \sigma_{r\theta} &= \mu\left[\frac{1}{r}\frac{\partial u_r}{\partial \theta} + r\frac{\partial}{\partial r}\left(\frac{u_\theta}{r}\right)\right] \\[2mm] \sigma_{rz} &= \mu\left[\frac{\partial u_r}{\partial z} + \frac{\partial u_z}{\partial r}\right] \end{aligned} \right\} . \qquad (3.41)$$

and

If we now consider the propagation of an infinite train of sinusoidal waves along a solid cylinder such that the displacement at each point is a simple harmonic function of z as well as

t we may write:

$$u_r = U \exp[i(\gamma z + pt)]$$
$$u_\theta = V \exp[i(\gamma z + pt)] \Bigg\},$$
$$u_z = W \exp[i(\gamma z + pt)]$$
(3.42)

where U, V, and W are functions of r and θ.

The frequency of the waves is $p/2\pi$, and their wavelength $2\pi/\gamma$, so that the phase velocity is given by p/γ. By substituting (3.42) in the equations (3.35), (3.36), and (3.37) and inserting the boundary conditions that the three stress components (3.41) vanish at the surface of the cylinder ($r = a$), expressions may be obtained for the phase velocity of waves of a given frequency.

We shall now consider the three types of vibration of cylindrical bars, namely, longitudinal, torsional, and lateral, in terms of these equations.

(a) Longitudinal waves

We assume here that the displacement u_θ vanishes, i.e. that each particle vibrates in its rz plane, and, further, that the motion is symmetrical about the axis of the cylinder so that U and W are independent of θ. Thus, from (3.39), $\bar{\omega}_r$ and $\bar{\omega}_z$ are both zero and equations (3.35) and (3.37) become:

$$-\rho p^2 u_r = (\lambda + 2\mu)\frac{\partial\Delta}{\partial r} + i2\mu\gamma\bar{\omega}_\theta,$$
(3.43)

$$-\rho p^2 u_z = i(\lambda + 2\mu)\gamma\Delta - \frac{2\mu}{r}\frac{\partial}{\partial r}(r\bar{\omega}_\theta)$$
(3.44)

(since, from (3.42), $\partial u_r/\partial t = ipu_r$, $\partial u_r/\partial z = i\gamma u_r$, etc.).

Using (3.38) and (3.39), we can eliminate either $\bar{\omega}_\theta$ or Δ from (3.43) and (3.44). This leads to the two equations:

$$\frac{\partial^2\Delta}{\partial r^2} + \frac{1}{r}\frac{\partial\Delta}{\partial r} + h'^2\Delta = 0$$
(3.45)

and

$$\frac{\partial^2\bar{\omega}_\theta}{\partial r^2} + \frac{1}{r}\frac{\partial\bar{\omega}_\theta}{\partial r} - \frac{\bar{\omega}_\theta}{r^2} + \kappa'^2\bar{\omega}_\theta = 0,$$
(3.46)

where

$$h'^2 = \rho p^2/(\lambda + 2\mu) - \gamma^2$$
(3.47)

and

$$\kappa'^2 = \rho p^2/\mu - \gamma^2.$$
(3.48)

If we change the variable in (3.45) from r to $h'r$, it becomes Bessel's equation of zero order, so that the solution which is finite at the axis is:

$$\Delta = G J_0(h'r). \qquad (3.49)$$

Similarly, taking $\kappa'r$ as the variable in (3.46), it becomes Bessel's equation of order one, so that:

$$\bar{\omega}_\theta = H J_1(\kappa'r), \qquad (3.50)$$

where G and H are functions of z and t, but are independent of r.

Now, substituting for u_r and u_z from (3.42) in (3.38) and (3.39), we have:

$$\Delta = \left[\frac{\partial U}{\partial r} + \frac{U}{r} + i\gamma W\right]\exp[i(\gamma z + pt)] \qquad (3.51)$$

and

$$2\bar{\omega}_\theta = \left[i\gamma U - \frac{\partial W}{\partial r}\right]\exp[i(\gamma z + pt)]. \qquad (3.52)$$

In order to satisfy (3.49), (3.50), (3.51), and (3.52), U and W must have the form:

$$U = A\,\frac{\partial}{\partial r}J_0(h'r) + C\gamma J_1(\kappa'r) \qquad (3.53)$$

and

$$W = Ai\gamma J_0(h'r) + \frac{Ci}{r}\frac{\partial}{\partial r}[rJ_1(\kappa'r)], \qquad (3.54)$$

where A and C are constants.

If we now substitute from (3.53) and (3.54) in the expressions for σ_{rr} and σ_{rz} given by (3.41), the condition that both these stress components vanish at the surface of the cylinder (where $r = a$) becomes:

$$A\left[2\mu\,\frac{\partial^2}{\partial a^2}J_0(h'a) - \frac{\lambda}{\lambda+2\mu}p^2\rho J_0(h'a)\right] + 2C\mu\gamma\,\frac{\partial}{\partial a}J_1(\kappa'a) = 0 \qquad (3.55)$$

and

$$2A\gamma\,\frac{\partial}{\partial a}J_0(h'a) + C\left(2\gamma^2 - \frac{p^2\rho}{\mu}\right)J_1(\kappa'a) = 0 \qquad (3.56)$$

(for brevity $\partial(\ )/\partial a$ is written here for $[\partial(\ )/\partial r]_{r=a}$). On eliminating A/C from (3.55) and (3.56), we obtain an equation involving the frequency and wavelength of the vibrations, the radius of the cylinder a, the elastic constants λ and μ, and the density ρ; this is known as the *frequency equation*. From this equation the phase velocity for sinusoidal waves of any frequency along an infinitely long cylinder may be obtained. These solutions are not exact, however, for a cylinder of finite length, since the conditions that the ends are free from traction cannot be obeyed by solutions of this type. When, however, the length of the cylinder is large compared with a, the residual stresses become very small.

If we expand $J_0(h'a)$ and $J_1(\kappa'a)$ as a power series, we have:

$$J_0(h'a) = 1 - \tfrac{1}{4}(h'a)^2 + \tfrac{1}{64}(h'a)^4 - \dots$$

and
$$J_1(\kappa'a) = \tfrac{1}{2}(\kappa'a) - \tfrac{1}{16}(\kappa'a)^3 + \dots .$$

Now if the radius of the cylinder a is sufficiently small for $h'a$ and $\kappa'a$ to be small compared with unity (which from (3.47) and (3.48) is found to imply that the wavelength of the vibrations is large compared with the radius of the cylinder), we can approximate to the frequency equation by using the expansions of the Bessel functions and neglecting powers of $h'a$ and $\kappa'a$ higher than the first. We then have:

$$J_0(h'a) = 1, \qquad \frac{\partial}{\partial a} J_0(h'a) = \tfrac{1}{2}h'^2 a, \ \text{ etc.}$$

If we insert these approximate values into (3.55) and (3.56), the frequency equation becomes:

$$(2\gamma^2 - p^2\rho)\kappa'a\left(h'^2 + \frac{\lambda p^2 \rho}{\mu(\lambda + 2\mu)}\right) = 2\gamma^2 \kappa' a. \qquad (3.57)$$

Rejecting the solution $\kappa' = 0$, which corresponds to waves travelling with velocity $(\mu/\rho)^{\frac{1}{2}}$ (see equation (3.48)), and substituting for h' from (3.47) we find that

$$\frac{p^2}{\gamma^2} = \frac{\mu(3\lambda + 2\mu)}{(\lambda + \mu)\rho}. \qquad (3.58)$$

This expresses the phase velocity in terms of the elastic constants and the density of the cylinder, and since $\mu(3\lambda + 2\mu)/(\lambda + \mu)$ is equal to Young's modulus E (see equation (2.4)), this velocity is $\sqrt{(E/\rho)}$, which is that obtained from the elementary treatment as given by equation (3.4).

By taking terms involving a^2 into account in the expansions of the Bessel functions, a better approximation is obtained, and the solution of the frequency equation becomes:

$$\frac{p}{\gamma} = \left(\frac{E}{\rho}\right)^{\frac{1}{2}} (1 - \tfrac{1}{4}\nu^2\gamma^2a^2), \qquad (3.59)$$

where ν is Poisson's ratio and equal to $\tfrac{1}{2}\lambda/(\lambda+\mu)$. Equation (3.59) was also derived by Rayleigh (1894), p. 252, from considerations of the energy associated with the lateral motion of the bar. Equation (3.59) may be expressed in non-dimensional form as

$$\frac{c_p}{c_0} = 1 - \nu^2\pi^2\left(\frac{a}{\Lambda}\right)^2, \qquad (3.60)$$

where $c_p = p/\gamma$, the phase velocity of the waves, $c_0 = \sqrt{(E/\rho)}$, the velocity of infinitely long waves in a bar, and $\Lambda = 2\pi/\gamma$, the wavelength. Equation (3.60) shows that the phase velocity decreases with decreasing wavelength and leads to the result that the velocity is zero at a wavelength of $\nu\pi a$. The equation cannot therefore be expected to be reliable except for wavelengths which are long compared with the radius of the bar.

The group velocity c_g may be derived from (3.60) by using the relation (3.27): this leads to

$$\frac{c_g}{c_0} = 1 - 3\nu^2\pi^2\left(\frac{a}{\Lambda}\right)^2. \qquad (3.61)$$

The group velocity would thus appear to reach zero at a wavelength of $\pi\nu a\sqrt{3}$. The curves obtained from equations (3.60) and (3.61) for $\nu = 0.29$ (the value for steel) are shown as the broken curve 1A in Figs. 14 and 15.

Although the Pochhammer treatment described in this section was first published in 1876 and has been quite familiar to subsequent workers in the field (Lord Rayleigh refers to it in *The Theory of Sound* (1894), p. 252), it is only during the last few years that numerical results have been computed from the frequency equation derived from (3.55) and (3.56). The results for longitudinal waves have been given by Field (1931), Bancroft (1941), Czerlinsky (1942), Mindlin (1946), and Davies (1948), whilst numerical results, from a similar treatment of flexural vibrations, which will be discussed later, have been published by Hudson (1943).

As they stand, (3.55) and (3.56) lead to a frequency equation involving six parameters. These are the elastic constants λ and μ, the density ρ, the radius of the bar a, the frequency of the waves $p/2\pi$, and their wavelength, $2\pi/\gamma$. By expressing this equation in non-dimensional form, however, the number of variables may be reduced to three, these being (c_p/c_0), (a/Λ), and ν. Thus, for any given value of ν, an equation involving only (c_p/c_0) and (a/Λ) is obtained. This equation is found to have multiple roots, each root corresponding to a different mode of vibration of the bar. Curves 1, 2, and 3 in Fig. 14 show the values calculated by Davies (1948) for the first three roots of the frequency equation, taking $\nu = 0\cdot29$. Bancroft (1941), on the other hand, has calculated the first branch of the curve for a series of values of ν, and the curve interpolated at $\nu = 0\cdot29$ from his results agrees with that obtained by Davies.

In Fig. 14 the values of c_1/c_0, c_2/c_0, and c_s/c_0 are also shown, c_1 and c_2 being the velocities of dilatational and distortional waves in an infinite medium and c_s the velocity of Rayleigh surface waves on a semi-infinite medium. These three ratios may be expressed in terms of ν. Thus:

$$\frac{c_1^2}{c_0^2} = \frac{\lambda+2\mu}{E} = \frac{(1-\nu)}{(1+\nu)(1-2\nu)}, \tag{3.62}$$

$$\frac{c_2^2}{c_0^2} = \frac{\mu}{E} = \frac{1}{2(1+\nu)}, \tag{3.63}$$

whilst c_s/c_2 is given by the cubic equation (2.37) of the previous chapter and c_s/c_0 may then be derived from (3.63). For $\nu = 0\cdot29$, $c_s/c_2 = 0\cdot9528$ so that $c_s/c_0 = 0\cdot5764$.

It may be seen from curve 1 in Fig. 14 that for long wavelengths, $a/\Lambda < 0\cdot1$, the phase velocity of longitudinal waves differs very little from c_0 and the Rayleigh correction (equation 3.60), shown by curve 1A, describes the dependence of wavelength on frequency quite adequately for this mode of vibration. For shorter wavelengths the errors become more serious, but curve 1A continues to afford a useful approximation for values of a/Λ up to about 0·7. At higher values than this curves 1 and 1A diverge rapidly, and whilst the exact theory predicts that the phase velocity will approach c_s asymptotically at very short wavelengths (Bancroft, 1941, has shown that the frequency equation reduces to the cubic for surface waves, equation

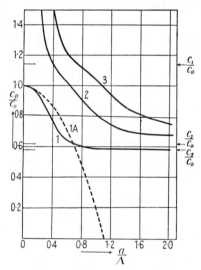

Fig. 14. Phase velocity of extensional waves
in cylindrical bars for $\nu = 0·29$.

(2.34), when a/Λ is very large), curve 1A cuts the a/Λ axis at 1·098.

In order to determine the distribution of displacement and stress across a section of the bar corresponding to a given root of the frequency equation, the value of the ratio A/C given by this root must first be obtained from equations (3.55) and (3.56). By inserting this value of A/C in (3.53) and (3.54), expressions for U and W are derived which involve only one constant; this is determined by the amplitude of the waves. Equations (3.42) give u_r and u_z in terms of U and W, and the corresponding stress components may then be obtained from equations (3.41) By this method Davies (1948) has calculated the distribution of the displacements and stress components across a section of a cylindrical steel bar for the first root of the frequency equation.

These calculations show that the first mode, given as curve 1 in Fig. 14, corresponds to vibrations in which there is no nodal cylinder up to a certain value of a/Λ (in the case of $\nu = 0\cdot29$, the value is about $0\cdot375$). At this value a nodal cylinder appears at the surface of the bar, and for values of a/Λ greater than this the first mode of vibration involves one nodal cylinder. The second mode, curve 2 in Fig. 14, involves one or more nodal cylinders and so on. The particular mode of vibration will depend on the initial conditions, and it is found experimentally that it is the first mode which is usually excited. As might have been expected from the fact that when a/Λ is large the phase velocity in the first mode approaches that of surface waves, the longitudinal displacement u_z, under these conditions, is found to be very large at the surface of the cylinder and to fall off rapidly with depth, the waves being analogous to Rayleigh waves on the surface layers of a semi-infinite medium.

It may be seen from Fig. 14 that the phase velocities for the higher modes exceed the velocity of dilatational waves in the medium. As mentioned earlier, however, the phase velocity does not correspond to the transmission of energy; this is propagated at the group velocity c_g. Fig. 15 shows the c_g curves corresponding to branches 1 and 2 of the phase velocity curves in Fig. 14. These are derived by differentiating the corresponding c_p curves and using relation (3.27). The values of c_g obtained from equation (3.61), using the Rayleigh correction, are also shown as the broken curve 1A. It may be seen that here again curve 1A is only reliable at very small values of a/Λ. Curve 2 shows that for this mode the group velocity, unlike the phase velocity, does not exceed c_0. It may also be seen that in the first mode the group velocity reaches a minimum when a/Λ equals about $0\cdot45$, whilst the second mode passes through a maximum at about the same value. This implies that when a pulse is propagated along a steel bar in the first mode, Fourier components of wavelengths about $0\cdot45$ times the radius of the bar will be found at the tail of the pulse.

The foregoing treatment of the propagation of extensional harmonic waves along an infinite cylinder in terms of the exact elastic equation leads to the result that energy cannot be transmitted along a cylinder by this type of wave at a velocity greater than c_0. Several writers, Field (1931), Southwell (1941), p. 355, Prescott (1942), and Cooper (1947) have pointed out,

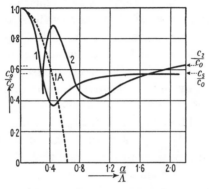

FIG. 15. Group velocity of extensional waves in cylindrical bars for $\nu = 0.29$.

however, that in theory one should be able to treat a cylinder in the same way as an infinite medium. Elastic waves would then be expected to travel with the only two possible velocities for an infinite medium (c_1 and c_2), and these waves will be continually reflected at the free surface of the cylinder in the way described in the previous chapter. Thus, if we consider a disturbance at a point in the interior of the cylinder, a spherical dilatation wave will travel out from it with velocity c_1, and some part of this wave will travel down the cylinder without suffering any reflections at the surface. The amplitude of this unreflected wave will decrease inversely with the distance, so that its effect will rapidly fall off, but some energy will nevertheless have been transmitted with the dilatation velocity in the medium. The parts of the wave that impinge on the cylindrical surface will give rise to reflected dilatation and distortion waves, and these in turn will give rise to waves of both types when they are again reflected. It is thus reasonable to expect that the greater part of the energy of the disturbance will travel with a velocity less

than that of dilatation waves. The Pochhammer theory, however, predicts that *no* energy can be transmitted at a velocity
greater than c_0, and experimental evidence seems to be required
to resolve this paradox.

It should be pointed out that the Pochhammer treatment
deals with sinusoidal waves travelling along an infinite cylinder
and, as shown by Love (1927), p. 288, equations (3.35), (3.36),
and (3.37) cannot be satisfied for the free vibrations of a cylinder
of finite length by harmonic solutions of the type (3.42), if it is
assumed that the ends of the cylinder are free from stress. For
an infinite cylinder, on the other hand, the fraction of the energy
communicated by spherical wavelets will eventually fall to
zero, so that the arguments of the preceding paragraph will not
apply. It should here be noted that the Pochhammer equations
are no more than the equations of motion of an elastic medium
referred to cylindrical coordinates, and that, if these equations
are applied to an unbounded medium, they will show two, and
only two, types of wave travelling separately with velocities
c_1 and c_2.

It seems clear that if a force is suddenly applied to the centre
of one face of a cylindrical disk whose thickness is comparable
to its radius, some energy will reach the other face at the dilatational velocity. Whether the same will apply when a uniform
plane wave impinges on the plate is more uncertain, and may
well depend on the end conditions of the plate. Bancroft (1941)
suggests that for plane waves we must distinguish between the
case where lateral motion is allowed and where it is inhibited,
either by the application of external forces, or by the presence
of the undisturbed medium. Where lateral motion can occur
freely, the operative elastic constant will be E, whilst where
it is forbidden it will be $(\lambda+2\mu)$.

Before leaving the theory of the propagation of extensional
waves in bars, it is perhaps worth referring to the treatment
due to Giebe and Blechschmidt (1933), since it was on the basis
of this theory that much of the subsequent experimental work
in Germany and America was considered. According to this
theory, a vibrating bar may be treated as two separate mechani-

cal systems, each with its own set of resonant frequencies. The observed resonant frequencies of the bar are considered to arise from the coupling between these two mechanical systems. For a cylindrical bar, the first set of resonant frequencies are taken to be those of a bar of infinitesimal cross-section in longitudinal vibrations, whilst the second set are those of a disk of infinitesimal thickness in radial vibration. Giebe and Blechschmidt assumed that only the fundamental frequency of radial vibration would be excited and combined this with the various possible longitudinal modes.

The curves they obtained for the dependence of phase velocity on wavelength show two branches, one of which is somewhat similar to curve 1A in Fig. 14, whilst the other is similar to curve 2 in the same figure, but displaced to the right. The most striking prediction of this theory is that there is a 'dead zone' between these two branches which corresponds to a frequency region in which it is impossible to excite longitudinal vibrations in a bar. Giebe and Blechschmidt claim to have found such a region experimentally for thin-walled cylindrical tubes to which they have also applied their theory. There is no evidence, however, for such a dead zone with solid cylinders where the experimental results are found to be in excellent agreement with curve 1 given in Fig. 14.

The Giebe and Blechschmidt treatment is found to be a better approximation to the truth than the Rayleigh correction, and provides an interesting physical model of the propagation of extensional waves in a bar. It has, however, now been superseded, at least so far as solid cylinders are concerned, by the results obtained from the exact elastic equations.

(b) Torsional waves

For the propagation of torsional waves we must find a solution of the equations of motion (3.35), (3.36), and (3.37), for which there are no longitudinal or lateral displacements, and the motion is symmetrical about the axis of the cylinder; i.e. u_r and u_z must both vanish and u_θ must be independent of θ. Inserting these conditions in (3.38), it may be seen that the dilatation Δ is zero, whilst from (3.39) the components of the rotation become:

$$2\bar{\omega}_r = -\frac{\partial u_\theta}{\partial z}, \qquad 2\bar{\omega}_z = \frac{u_\theta}{r} + \frac{\partial u_\theta}{\partial r}, \quad \text{and} \quad \bar{\omega}_\theta = 0.$$

The equations of motion (3.35) and (3.37) are now satisfied identically, whilst (3.36) becomes:

$$\frac{\rho}{\mu}\frac{\partial^2 u_\theta}{\partial t^2} = \frac{\partial^2 u_\theta}{\partial z^2} - \frac{u_\theta}{r^2} + \frac{1}{r}\frac{\partial u_\theta}{\partial r} + \frac{\partial^2 u_\theta}{\partial r^2}. \tag{3.64}$$

If we now consider harmonic waves and take the expression for u_θ from (3.42), equation (3.64) becomes:

$$\frac{\partial^2 V}{\partial r^2} + \frac{1}{r}\frac{\partial V}{\partial r} + \left(\frac{\rho p^2}{\mu} - \gamma^2 - \frac{1}{r^2}\right)V = 0. \tag{3.65}$$

If, as in equation (3.48), we write κ'^2 for $(\rho p^2/\mu - \gamma^2)$ and change the variable from r to $\kappa'r$ (3.65) becomes Bessel's equation of order one, and the solution which is finite at the axis is:

$$V = BJ_1(\kappa'r), \quad \text{where } B \text{ is a constant.} \tag{3.66}$$

The condition that the three stress components σ_{rr}, $\sigma_{r\theta}$, and σ_{rz} vanish at the surface of the cylinder leads to only one equation, since from (3.41) it will be seen that σ_{rr} and σ_{rz} are now zero everywhere.

From (3.66) the condition that $\sigma_{r\theta}$ is zero when r equals a becomes:

$$\frac{\partial}{\partial a}\left[\frac{J_1(\kappa'a)}{a}\right] = 0. \tag{3.67}$$

This equation has multiple roots, the first of which is $\kappa' = 0$. This value of κ' cannot, however, be inserted in (3.66), since (3.65) can only be treated as a Bessel equation when κ' is not zero. We must therefore go back to (3.65) and try to insert this value of κ', and we then find that

$$V = B'r \tag{3.68}$$

satisfies the equation, where B' is a constant. From (3.41) it may be seen that this expression for V makes $\sigma_{r\theta}$ vanish, so that the boundary conditions at the surface of the cylinder are satisfied. Since the amplitude of u_θ is proportional to r, and u_r and u_z are both zero, the motion corresponding to this solution is a rotation of each transverse section of the cylinder as a whole about its centre. Now $\kappa'^2 = \rho p^2/\mu - \gamma^2$, so that the phase velocity corresponding to $\kappa' = 0$ is given by:

$$\frac{p}{\gamma} = \sqrt{\left(\frac{\mu}{\rho}\right)}. \tag{3.69}$$

There is thus no dispersion for waves of this type, and the phase velocity and group velocity are both equal to the velocity of distortion waves in an infinite medium. This result is in agreement with that obtained from the simple treatment as given by equation (3.18).

If we differentiate equation (3.67) by parts and use the recurrence formulae for Bessel functions (see for example Whittaker and Watson, 1946, p. 359) we obtain:

$$\frac{\kappa' a J_0(\kappa' a)}{J_1(\kappa' a)} = 2. \tag{3.70}$$

The roots of this equation may be determined numerically from tables of Bessel functions. They give a series of values of $\kappa' a$ corresponding to more complex modes of torsional vibration which involve nodal cylinders. If we denote one of these roots by K_1 we have:

$$K_1^2 = a^2\left(\frac{p^2\rho}{\mu} - \gamma^2\right). \tag{3.71}$$

Now $\sqrt{(\mu/\rho)}$ is equal to c_2, the velocity of waves of distortion in an infinite medium, whilst p/f is the phase velocity of the torsion waves along the cylinder, which will be denoted by c_t, and f is equal to $2\pi/\Lambda$, where Λ is the wavelength of these torsion waves. We may thus write (3.71) in non-dimensional form as:

$$K_1^2 = 4\pi^2\left(\frac{a}{\Lambda}\right)^2\left(\frac{c_t^2}{c_2^2} - 1\right)$$

or
$$\frac{c_t}{c_2} = \left(K_1^2\frac{(\Lambda/a)^2}{4\pi^2} + 1\right)^{\frac{1}{2}}. \tag{3.72}$$

Equation (3.72) gives the relation between the phase velocity and the wavelength of torsional waves for the various values of K_1. It may be seen that in all these modes dispersion takes place, the phase velocity becoming infinite for very long wavelengths and equal to c_2 for very short wavelengths. By differentiating (3.72) with respect to Λ and substituting in equation (3.27), the expression for the group velocity of these torsional waves may be determined; this velocity comes out as c_2^2/c_t, so that the group velocity of these torsional waves starts from zero for very long wavelengths and approaches c_2 asymptotically for very short wavelengths.

(c) Flexural waves

For longitudinal and torsional waves travelling along a cylindrical bar the motion is symmetrical about the axis of the bar, so that the displacements are independent of θ and the amplitude functions U, V, and W in equation (3.42) are simply functions of r. The treatment of these types of wave motion is also simplified by the fact that for longitudinal waves V is zero, whilst for torsional motion both U and W are zero.

For flexural waves, however, all three components of the displacement must be considered, and all three involve θ. The treatment in terms of the Pochhammer equations consequently becomes very complex, and will not be given in detail here. An outline of it will be found in Love (1927), p. 291, and the resulting frequency equation is discussed by Bancroft (1941), and by Hudson (1943), who treats the general case of vibrations of cylinders and derives the frequency equation for flexural waves as a special case.

If we take the undisturbed axis of the bar as the z-axis and assume that vibrations take place in the plane containing this axis and the line from which θ is measured it is reasonable to try solutions in which u_r and u_z are proportional to $\cos\theta$, and u_θ is proportional to $\sin\theta$. We may then write instead of equations (3.42):

$$\left. \begin{aligned} u_r &= U'\cos\theta\exp[i(\gamma z+pt)] \\ u_\theta &= V'\sin\theta\exp[i(\gamma z+pt)] \\ u_z &= W'\cos\theta\exp[i(\gamma z+pt)] \end{aligned} \right\}, \qquad (3.73)$$

where U', V', and W' are functions of r only.

For the sake of clarity, we shall call the plane from which θ is measured 'vertical' and the direction of the z-axis 'horizontal'. Thus θ is zero for points in the vertical section of the bar which contains the z-axis. The second of equations (3.73) implies that for points in this plane u_θ is zero, so that these points will remain in the plane during the vibration. For points in the horizontal plane containing the axis of the bar (this plane corresponds to the 'neutral axis' in the elementary treatment of flexural vibrations), θ is $\frac{1}{2}\pi$, hence from the first and third equations in (3.73), u_r and u_z are both zero. Points in this plane will consequently perform purely vertical oscillations, since the only remaining component of the displacement u_θ is perpendicular to both the radius vector r and the z-axis.

Thus equations (3.73) appear to correspond to motion of lateral or flexural type taking place in the plane from which θ is measured, and it remains to determine if expressions can be found for U', V', and W' which will satisfy the equations of motion (3.35), (3.36), and (3.37), and also the boundary conditions at the surface of the bar.

It is found that the following expressions will satisfy the equations of motion:

$$
\left.
\begin{aligned}
U' &= A\frac{\partial}{\partial r}J_1(h'r) + B\gamma\frac{\partial}{\partial r}J_1(\kappa'r) + Cr^{-1}J_1(\kappa'r) \\
V' &= Ar^{-1}J_1(h'r) - B\gamma r^{-1}J_1(\kappa'r) - C\frac{\partial}{\partial r}J_1(\kappa'r) \\
W' &= Ai\gamma J_1(h'r) - Bi\kappa'^2 J_1(\kappa'r)
\end{aligned}
\right\}, \quad (3.74)
$$

where A, B, and C are constants, and h' and κ' are given by equations (3.47) and (3.48) respectively. (The arguments by which the above expressions are derived are described by Love (1927), p. 291, and are similar to those used earlier in this section to obtain the expressions U and W given in equations (3.53) and (3.54). It may be shown that the equations of motion lead to Bessel equations of order one for the dilatation Δ and for the component of the rotation $\bar{\omega}_z$. It is hence found that Δ is proportional to $J_1(h'r)$, whilst $\bar{\omega}_z$ is proportional to $J_1(\kappa'r)$, and it may then be shown that $\bar{\omega}_r$ must be proportional to

$$
Gp^2\rho\mu^{-1}r^{-1}J_1(\kappa'r) + H\gamma\frac{\partial}{\partial r}[J_1(\kappa'r)],
$$

where G and H are constants. In order to satisfy these relations for Δ, $\bar{\omega}_z$, and $\bar{\omega}_r$ by expressions (3.73), it is then found that U', V', and W' must be of the form given by (3.74).)

In order to insert the boundary conditions, equations (3.74) must be used in the expressions (3.41) for σ_{rr}, $\sigma_{r\theta}$, and σ_{rz}, and these three stress components must be zero at the surface of the cylinder where $r = a$. This leads to three simultaneous equations from which A, B, and C can be eliminated (these equations can be written to involve only two constants, e.g. A/B and A/C), and the frequency equation may hence be derived. Love states that, as in the case of longitudinal waves, the solutions will not apply exactly to the free flexural vibrations of a cylinder of finite length since the condition that the end faces are free from stress cannot be satisfied exactly by these solutions. When the length of cylinder is large compared with its radius however, the residual stresses will be very small.

The frequency equation for flexural waves is given in determinantal form by Bancroft (1941), who suggested that it was unlikely to be useful owing to its complexity. Hudson (1943), however, has managed to carry out the necessary computations from this equation and has obtained values which show how the phase velocity of flexural waves depends on the ratio between their wavelength and the radius of the cylinder. He has also shown that for flexural waves the frequency equation has only one root, so that these waves, unlike their extensional and torsional counterparts, can be propagated in only one mode.

Hudson gives his results in non-dimensional form as a table of the ratio of the phase velocity of flexural waves in a bar to the velocity of distortion waves in an unbounded medium. These velocities are given for different values of the ratio between the wavelength of distortion waves and the circumference of the bar, and results are shown for a series of values of Poisson's ratio ν. Davies (1948) has interpolated from these values for the case of ν equal to 0·29, and the curves shown in Fig. 16 were obtained from his paper.

This figure is plotted in the same way as Fig. 14, and gives the ratio between the phase velocities of flexural waves and of extensional waves of infinitely long wavelength. (The latter velocity is denoted by c_0 as before, and is equal to $(E/\rho)^{\frac{1}{2}}$.) This ratio is plotted against a/Λ, where a is the radius of the bar and

Λ is the wavelength of the flexural waves. For comparison the curves obtained from the elementary treatment (equation 3.26) and from the Rayleigh correction (equation 3.32), as well as the curve for the first mode of extensional waves from Fig. 14, are

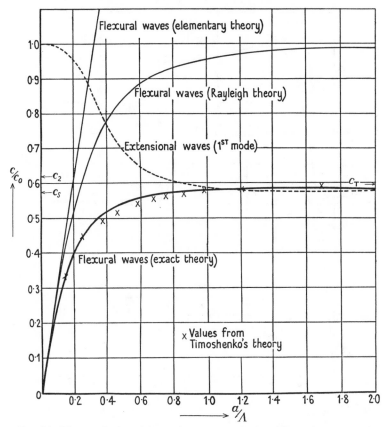

FIG. 16. Phase velocity of flexural waves in cylindrical bars (for $\nu = 0.29$).

shown in Fig. 16. The results obtained from Timoshenko's equation (3.34) are given as a series of points, and these may be seen to be in remarkably good agreement with those obtained by Hudson from the exact elastic equations. The discrepancy between these two treatments is greatest for large values of a/Λ, but even for the limiting case of waves of infinitely short wavelength the difference is not large. The exact theory predicts that the limiting velocity will be that of Rayleigh surface waves, c_s, which for $\nu = 0.29$ comes out as $0.5764c_0$, whilst Timoshenko's treatment gives the limiting velocity as $0.5906c_0$; this value is denoted as c_T in the figure.

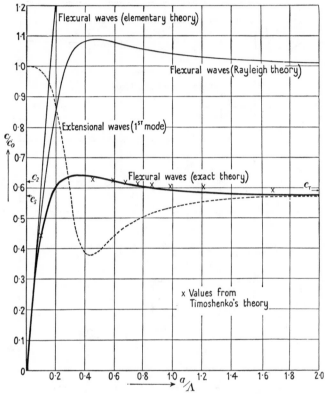

FIG. 17. Group velocity of flexural waves in cylindrical bars (for $\nu = 0.29$).

It may be seen from Fig. 16 that for wavelengths greater than about ten times the radius of the bar, all four treatments give sensibly the same results. For wavelengths less than this, however, both the elementary theory and the Rayleigh correction lead to large errors.

Fig. 17 shows the group velocities derived from the curves given in Fig. 16, and the most interesting feature here is that the exact theory predicts that the group velocity will have a maximum at a/Λ equal to about 0·3. This implies that when a flexural pulse is propagated along a bar, Fourier components whose wavelengths are about three times that of the radius of the bar will travel ahead of components of other wavelengths, and will therefore appear at the head of the pulse.

Propagation of an elastic pulse along a cylindrical bar

The elementary theory of the propagation of elastic disturbances along a cylindrical bar predicts that dispersion will occur with flexural pulses, whilst longitudinal and torsional pulses will be propagated along a bar without change in form.

The Pochhammer treatment described in the last section shows that the velocity of propagation of longitudinal sinusoidal waves depends on their wavelength, and it is only when torsional waves are travelling in their fundamental mode that dispersion is absent. This theory also shows that for all three types of waves the elementary theory applies only when the wavelength is large compared with the radius of the bar. The results of the exact theory cannot be readily applied to the propagation of a single pulse, since such a pulse can only be analysed into sinusoidal components in terms of a Fourier integral, which is generally intractable. The type of distortion produced in a pulse as it travels down the bar can, however, be estimated from the dispersion curves shown in Figs. 14–17.

Davies (1948) has treated the case of a longitudinal pulse in terms of the exact theory in two ways. One method he has employed is to consider the propagation of a periodically repeating pulse. Such a repeating pulse can be analysed into a Fourier series and the velocity of propagation appropriate to each term of the series may be found from the curve for phase velocities. The second treatment used was Kelvin's method of stationary phase. This method treats the propagation of an infinitely thin pulse of infinitely large amplitude. Such a pulse can be expressed as a Fourier integral, and may be regarded as having been produced by the superposition of sinusoidal stress waves covering a spectrum of wavelengths. All the wave trains are taken to be of equal amplitude, and considered to be in phase at the origin and to cancel out everywhere else, at time $t = 0$. The stress distribution at any subsequent time may then be investigated from the curves of group velocity.

By these methods Davies has shown that a longitudinal pulse whose original length is comparable with the radius will become distorted on travelling down the bar, and the main pulse will be followed by a train of oscillations of high frequency; further, any sharp changes in gradient will be rounded off, and straight parts of the pulse will become oscillatory curves. He has confirmed these results experimentally, and has shown that they may be predicted from the approximate equation for longitudinal waves which allows for the effects of lateral inertia, and is given by Love (1927), p. 428.

Propagation along bars of non-circular cross-section

The elementary theory of the propagation of elastic waves along cylindrical bars described at the beginning of this chapter may be extended to bars of any uniform cross-section so long as the wavelength is large compared with the lateral dimensions of the bar. According to these treatments, longitudinal waves are propagated with constant velocity $c_0 = (E/\rho)^{\frac{1}{2}}$, the velocity of torsional waves will depend on the shape of the cross-section but will be constant for any one shape. Flexural waves, however, are dispersed, the phase velocity of sinusoidal flexural waves of wavelength Λ being given by $2\pi K c_0/\Lambda$, where K is the radius of gyration of a cross-section of the bar about an axis in the neutral surface perpendicular to the axis of the bar (see equation (3.26)). When the wavelengths become comparable with the lateral dimensions of the bar these relations break down, and the exact elastic equations must be used to investigate the nature of the propagation. The exact treatment for cylindrical bars has been discussed in the previous section, but for bars of non-circular cross-section the analysis becomes exceedingly complex and only in a few cases have solutions been attempted.

Chree (1889) discusses the propagation of longitudinal waves along bars of elliptical and rectangular cross-sections and derives the following approximate expression for the phase velocity

c_p of waves of wavelength $2\pi/\gamma$:

$$c_p = c_0(1 - \tfrac{1}{2}\nu^2\gamma^2 I_1).$$ (3.75)

This is a generalized form of Rayleigh's equation (3.59) which was derived for cylindrical bars. I_1 is here the second moment of the cross-section about the axis of the bar, and will be equal to $\tfrac{1}{4}(a^2+b^2)$ for a bar of elliptic cross-section, where a and b are the major and minor axes of the ellipse, whilst for a rectangular bar I_1 is equal to $\tfrac{1}{3}(a^2+b^2)$, where a and b are the sides of the rectangular sections. Equation (3.75) is accurate if the greatest lateral dimension of the bar a' is sufficiently small compared with the wavelength Λ for terms in (a'/Λ) of degree higher than the second to be ignored.

Morse (1950) also has considered the propagation of longitudinal waves along bars of rectangular cross-section in terms of the exact elastic equations, and has obtained solutions for bars whose width is large compared with their thickness. He has shown that these solutions are in good agreement with his experimental results (Morse, 1948). The experimental results lie on two separate curves corresponding to the two branches 1 and 2 of the theoretical curves for cylinders shown in Fig. 14.

Morse has found experimentally that a square bar gives a dispersion curve indistinguishable from that obtained with a cylindrical bar, if the ratio of the diameter of the cylinder to the side of the square cross-section is 1·13. This was found to hold over the whole dispersion curve. According to Chree's expression, equation (3.75), the ratio should be 1·15, and it is hard to say whether this small difference is due to experimental error or to the approximate nature of equation (3.75).

Propagation along a conical bar

Longitudinal harmonic waves whose wavelength is large compared with the lateral dimensions of a bar are propagated along it with constant velocity c_0. A pulse composed of such waves will thus travel without change of form down a bar of uniform cross-section at this velocity. If, however, the cross-section is not uniform along the length of the bar the shape of a pulse as well as its amplitude changes during its passage.

The propagation of a longitudinal pulse along a conical bar has been considered by Landon and Quinney (1923) who employed such a bar to measure the pressures produced by the detonation of explosives. The experimental method these workers employed was due originally to Hopkinson (1914) and will be described in the next chapter. The approximate theory

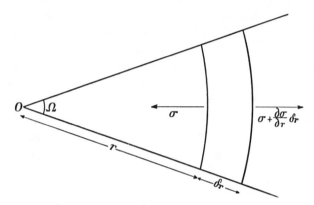

Fig. 18. Stress distribution in a conical bar.

of the propagation of longitudinal waves along a conical bar of small apical angle will, however, be discussed here as this treatment illustrates certain differences between the behaviour of plane and spherical stress waves and the results are also relevant to some experiments on the fracture of cones described in Chapter VIII. This treatment will apply only to waves whose wavelength is large compared with the diameter of the cone in the region of the bar in which they are propagated.

We take the cone to be of small solid angle Ω with its apex at the origin O (see Fig. 18). We now consider the forces acting on an element of the cone bounded by spheres of radius r and $r + \delta r$. If Ω is sufficiently small, the normal stress will be uniform for each of the two spherical surfaces, and we denote this stress by σ over the surface of radius r, and hence by $\sigma + \dfrac{\partial \sigma}{\partial r} \delta r$ on the surface of radius $(r + \delta r)$. Now, these normal stresses will be acting everywhere in a direction almost parallel to the axis of the cone, so that the equation of motion in this direction may be written:

$$\rho r^2 \Omega \, \delta r \frac{\partial^2 u_1}{\partial t^2} = \left(\sigma + \frac{\partial \sigma}{\partial r} \delta r\right)(r+\delta r)^2 \Omega - \sigma r^2 \Omega, \qquad (3.76)$$

where ρ is the density of the cone and u_1 is the displacement parallel to the direction of the axis of the cone. Multiplying out, and neglecting the terms involving $(\delta r)^2$, (3.76) becomes:

$$r\frac{\partial \sigma}{\partial r} + 2\sigma = \rho r \frac{\partial^2 u_1}{\partial t^2}. \qquad (3.77)$$

Now the elastic relation for the element if it is considered as plane is $\sigma = E \dfrac{\partial u_1}{\partial r}$, where E is Young's modulus, so that (3.77) becomes:

$$\rho r \frac{\partial^2 u_1}{\partial t^2} = Er \frac{\partial^2 u_1}{\partial r^2} + 2E \frac{\partial u_1}{\partial r} \quad \text{or} \quad \frac{\partial^2 (u_1 r)}{\partial t^2} = c_0^2 \frac{\partial^2 (u_1 r)}{\partial r^2}, \quad (3.78)$$

where $c_0^2 = E/\rho$.

Equation (3.78) is the wave equation for spherical waves, and the solution, as in equation (2.20) of the previous chapter, is

$$u_1 r = f(r - c_0 t) + F(r + c_0 t), \qquad (3.79)$$

where $f(\ )$ and $F(\ )$ represent waves travelling in opposite directions.

If we consider the wave represented by

$$u_1 = \frac{1}{r} F(r + c_0 t) \qquad (3.80)$$

we have for the stress

$$\sigma = E \frac{\partial u_1}{\partial r} = \frac{E}{r} F'(r + c_0 t) - \frac{E}{r^2} F(r + c_0 t), \qquad (3.81)$$

whilst the particle velocity is given by:

$$\frac{\partial u_1}{\partial t} = \frac{c_0}{r} F'(r + c_0 t). \qquad (3.82)$$

There is thus no simple relation between stress and particle velocity for these waves except when r is sufficiently large for the second term in (3.80) to be ignored in comparison with the first. We then have:

$$\sigma = \frac{E}{c_0}\frac{\partial u_1}{\partial t} = \rho c_0 \frac{\partial u_1}{\partial t}. \qquad (3.83)$$

This is the same relation as equation (3.8), which was derived for plane waves. Now (3.80) can represent a pulse travelling towards the apex of the cone if we take $t = 0$ to correspond to the arrival of the head of the pulse at the apex of the cone, t being negative whilst the pulse is travelling towards the apex.

$$u_1 = \frac{A}{r}\left[\exp\!\left(\frac{-r-c_0 t}{\Lambda}\right) - 1\right] \qquad (3.84)$$

represents the displacement associated with a simple type of pulse. This expression holds for negative values of t, where r is numerically greater than $c_0 t$. At the head of the pulse $r = -c_0 t$, and the displacement is zero, whilst values of r less than $|c_0 t|$ correspond to the undisturbed region of the cone where the displacement again vanishes. In expression (3.84) Λ is the *characteristic length* of the pulse, and it is only in regions of the conical bar where this is great compared with the diameter of the bar that the approximate theory will apply. A is a constant of the dimensions of length squared and gives the amplitude of the pulse.

From (3.81) we find that the stress associated with the pulse is:

$$\sigma = \frac{-AE}{\Lambda r}\exp\!\left(\frac{-r-c_0 t}{\Lambda}\right) + \frac{AE}{r^2}\left[1 - \exp\!\left(\frac{-r-c_0 t}{\Lambda}\right)\right] \qquad (3.85)$$

(σ was taken to represent a tensile stress so that negative values of σ in (3.85) correspond to compression).

It may be seen from (3.85) that where $r \gg |c_0 t|$, i.e. in regions of the bar far behind the head of the pulse, a tensile stress AE/r^2 remains. This tension is necessary for the momentum in the bar to be conserved. The momentum associated with the pulse is given by:

$$\int_{\infty}^{-c_0 t} \rho r^2 \Omega \frac{\partial u_1}{\partial t}\, dr = -\int_{\infty}^{-c_0 t} \rho r^2 \Omega \frac{A c_0}{\Lambda r}\exp\!\left(\frac{-r-c_0 t}{\Lambda}\right) dr.$$

On integrating and inserting the limits this becomes

$$-A\rho c_0\Omega(c_0 t-\Lambda).$$

The momentum thus continually decreases at the rate of $A\rho c_0^2\Omega = AE\Omega$ per unit time (it should be remembered that t is here negative, and is decreasing in magnitude as the pulse approaches the apex). This decrease in momentum is balanced by the residual tensile stress AE/r^2, which is acting over a surface of area $r^2\Omega$.

Returning now to expression (3.85), it may be seen that the pulse consists of two terms of opposite sign and that the pressure wave is followed by a wave of tension, and as the pulse approaches the apex, the region of compression becomes shorter and shorter. By equating (3.85) to zero, the value of r at the boundary between the compression and tension regions may be found. This gives:

$$\frac{1}{\Lambda}\exp\left(\frac{-r-c_0 t}{\Lambda}\right) = \frac{1}{r}\left[1-\exp\left(\frac{-r-c_0 t}{\Lambda}\right)\right].$$

Writing r_0 for $-c_0 t$, this may be expressed in non-dimensional form as:

$$\left(\frac{r}{\Lambda}+1\right)\exp\left(\frac{-r+r_0}{\Lambda}\right) = 1. \tag{3.86}$$

Now r_0 is the distance of the head of the pulse from the apex of the cone whilst r is the distance at which the compression wave ends. Hence the length of the compression wave is given by $(r-r_0)$ and (3.86) gives the length of the compression region for different values of r. These are as follows:

r_0/Λ .	.	8·0	4·0	2·0	1·0	0·5	0·25
$(r-r_0)/\Lambda$	.	2·44	1·94	1·50	1·15	0·85	0·63

Thus, as the head of the pulse approaches the apex, the compression region becomes shorter and shorter, and when it finally reaches the apex the conical bar is entirely under tension. The pulse is then reflected at the apex and in the region between the head of this reflected pulse and the apex of the cone the combined effects of the incident and reflected pulses must be considered.

Propagation of longitudinal waves in an infinite plate

Before leaving the theory of elastic waves in solids we shall discuss briefly the propagation of longitudinal waves in an infinite plate. This problem has been solved by Lamb (1917), who showed that for wavelengths small compared with the thickness of the plate the velocity of propagation became that of Rayleigh surface waves. When the wavelength is large compared with the thickness of the plate the stress is uniform over any cross-section of the plate perpendicular to the direction

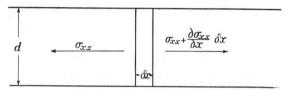

FIG. 19. Stress distribution in an infinite plate.

of propagation of the wave, and the equation of motion may be derived directly. Thus if we take the xy plane to be parallel to the surfaces of the plate and the propagation to be in the x-direction and consider an element of unit length in the direction of y and width δx (see Fig. 19), we have:

$$\rho d \frac{\partial^2 u}{\partial t^2} \delta x = \left(\sigma_{xx} + \frac{\partial \sigma_{xx}}{\partial x} \delta x\right) d - \sigma_{xx} d,$$

where d is the thickness of the plate and ρ is its density. Hence:

$$\frac{\partial \sigma_{xx}}{\partial x} = \rho \frac{\partial^2 u}{\partial t^2}. \tag{3.87}$$

To express σ_{xx} in terms of the displacement u, we must go back to the elastic relations (2.3), between stress and strain in an isotropic solid. With the directions chosen for the axes, v, the displacement in the y-direction, and σ_{zz}, the stress component perpendicular to the plate, will both be zero, so that we have from the first and third equations in (2.3)

$$\sigma_{xx} = (\lambda + 2\mu) \frac{\partial u}{\partial x} + \lambda \frac{\partial w}{\partial z},$$

$$\sigma_{zz} = (\lambda + 2\mu) \frac{\partial w}{\partial z} + \lambda \frac{\partial u}{\partial x} = 0.$$

Eliminating $\partial w/\partial z$ from these two equations we find

$$\sigma_{xx} = \frac{4\mu(\lambda+\mu)}{(\lambda+2\mu)}\frac{\partial u}{\partial x}. \tag{3.88}$$

Thus (3.87) becomes:

$$\rho\frac{\partial^2 u}{\partial t^2} = \frac{4\mu(\lambda+\mu)}{(\lambda+2\mu)}\frac{\partial^2 u}{\partial x^2}. \tag{3.89}$$

This is the wave equation and shows that the waves will be propagated with constant velocity given by:

$$c_L = \left[\frac{4\mu(\lambda+\mu)}{\rho(\lambda+2\mu)}\right]^{\frac{1}{2}}. \tag{3.90}$$

This velocity may be more conveniently expressed in terms of E and ν by substituting from equations (2.4) and (2.5); it then becomes:

$$c_L = \left[\frac{E}{\rho(1-\nu^2)}\right]^{\frac{1}{2}}. \tag{3.91}$$

This result will hold for wavelengths which are large compared with the thickness of the plate d. When the wavelength becomes comparable with the thickness, the stress distribution ceases to be uniform over a section of the plate perpendicular to the wave front. The exact elastic equations (2.8), (2.9), and (2.10) must then be used with the boundary condition that the surfaces of the plate are free from stress, the treatment being very similar to that described for Rayleigh waves in Chapter II. Lamb (1917) has considered the propagation of sinusoidal plane waves in an infinite plate, and has shown that where the motion is symmetrical about the medial plane of the plate the frequency equation may be written:

$$\frac{\tanh(\beta d/2)}{\tanh(\zeta d/2)} = \frac{4f^2\zeta\beta}{(f^2+\beta^2)^2}, \tag{3.92}$$

where $f = 2\pi$ divided by the wavelength, and β and ζ are two functions given by the relations:

$$\beta^2 = f^2\left(1-\frac{c^2}{c_2^2}\right), \qquad \zeta^2 = f^2\left(1-\frac{c^2}{c_1^2}\right). \tag{3.93}$$

c is here the phase velocity of the waves in the plate, and c_1 and c_2 are as before the velocities of dilatational and distortional waves in an infinite medium. It should be noted that when $c > c_2$, β is a pure imaginary, and that if this is put equal to $i\beta_r$ (3.92) may be written:

$$\frac{\tan(\beta_r d/2)}{\tanh(\zeta d/2)} = \frac{4f^2\zeta\beta_r}{(f^2+\beta^2)^2}.$$

When the wavelength is large compared with d it may be seen from (3.92) that βd and ζd become small, so that the arguments may be substituted for their hyperbolic tangents, (3.92) then becomes:

$$4f^2\zeta^2 = (f^2+\beta^2)^2. \qquad (3.94)$$

Substituting for β^2 and ζ^2 from (3.93) equation (3.94) becomes:

$$4f^4\left(1-\frac{c^2}{c_1^2}\right) = f^4\left(2-\frac{c^2}{c_2^2}\right)^2,$$

hence

$$c^2 = 4c_2^2\frac{(c_1^2-c_2^2)}{c_1^2}.$$

Now $c_1^2 = (\lambda+2\mu)/\rho$ and $c_2^2 = \mu/\rho$, so that:

$$c^2 = \frac{4\mu(\lambda+\mu)}{\rho(\lambda+2\mu)}$$

which is identical with (3.90), the result obtained from the elementary theory.

For very short waves, both βd and ζd become very large, and their hyperbolic tangents approach unity. (3.92) then simplifies to:

$$(f^2+\beta^2)^2 = 4f^2\zeta\beta.$$

Squaring both sides of the equation and substituting from (3.93), we have:

$$\left(2-\frac{c^2}{c_2^2}\right)^4 = 16\left(1-\frac{c^2}{c_1^2}\right)\left(1-\frac{c^2}{c_2^2}\right). \qquad (3.95)$$

If $\kappa_1 = c/c_2$, and $\alpha_1 = c_2/c_1$, (3.95) becomes, on multiplying out:

$$\kappa_1^6-8\kappa_1^4+(24-16\alpha_1^2)\kappa_1^2+(16\alpha_1^2-16) = 0. \qquad (3.96)$$

This is identical with equation (2.37) for Rayleigh surface waves, which was derived in Chapter II, and, since

$$\alpha_1^2 = \frac{c_2^2}{c_1^2} = \frac{\mu}{(\lambda+2\mu)} = \frac{(1-2\nu)}{(2-2\nu)},$$

(3.96) may be solved for any value of Poisson's ratio ν.

Plane longitudinal waves in an infinite plate thus travel with velocity $c_L = \{E/(\rho[1-\nu^2])\}^{\frac{1}{2}}$ when the wavelength Λ is very large compared with the thickness of the plate d, and with the velocity of Rayleigh surface waves when the wavelength is very small compared with the thickness. For wavelengths comparable

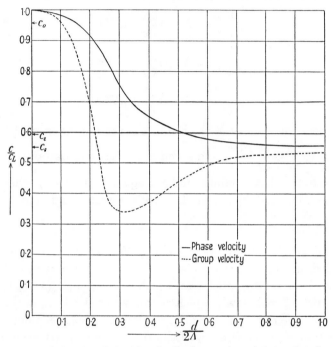

FIG. 20. Velocities of plane longitudinal waves in an infinite plate for $\nu = 0.29$.

with the thickness, dispersion takes place, the velocity depending on the ratio of wavelength to thickness. Fig. 20 shows the curves of phase velocity and group velocity plotted against $d/2\Lambda$ for $\nu = 0.29$. The phase velocity curve was computed from equation (3.92) and the group velocities were then obtained from relation (3.27); the velocities are plotted as the ratio c/c_L. It may be seen that the group velocity has a minimum value at d/Λ equal to about 0.6, and the shapes of the curves are rather similar to those found for the first mode in a cylindrical bar (cf. Figs. 14 and 15).

Summary of the theory of elastic waves in solids given in Chapters II and III

The equations of motion of an isotropic elastic solid in terms of the particle displacements u, v, and w which are in the x-, y-, and z-directions respectively are

$$\rho \frac{\partial^2 u}{\partial t^2} = (\lambda + \mu) \frac{\partial \Delta}{\partial x} + \mu \nabla^2 u, \qquad (2.8)$$

$$\rho \frac{\partial^2 v}{\partial t^2} = (\lambda + \mu) \frac{\partial \Delta}{\partial y} + \mu \nabla^2 v, \qquad (2.9)$$

and

$$\rho \frac{\partial^2 w}{\partial t^2} = (\lambda + \mu) \frac{\partial \Delta}{\partial z} + \mu \nabla^2 w. \qquad (2.10)$$

ρ is here the density of the solid, Δ is the dilatation, which is equal to $\frac{\partial u}{\partial x} + \frac{\partial v}{\partial y} + \frac{\partial w}{\partial z}$, and λ and μ are Lamé's constants. μ is equal to the shear modulus and λ is equal to $(k + 2\mu/3)$, where k is the bulk modulus. ∇^2 is Laplace's operator and stands for $\frac{\partial^2}{\partial x^2} + \frac{\partial^2}{\partial y^2} + \frac{\partial^2}{\partial z^2}$.

The solution of equations (2.8), (2.9), and (2.10) for an extended medium corresponds to two types of waves, called dilatational waves, which travel with velocity $c_1 = [(\lambda + 2\mu)/\rho]^{\frac{1}{2}}$, and distortional waves, which travel with velocity $c_2 = (\mu/\rho)^{\frac{1}{2}}$. The particle motion in the former is longitudinal, i.e. along the direction of propagation, whilst in the latter the particle motion is transverse, i.e. perpendicular to the direction of propagation.

Where there is a free surface, Rayleigh surface waves can also be propagated. These travel with a fraction κ_1 of the velocity of distortional waves c_2, the value of κ_1 being obtained from a cubic equation in κ_1^2:

$$\kappa_1^6 - 8\kappa_1^4 + (24 - 16\alpha_1^2)\kappa_1^2 + 16\alpha_1^2 - 16 = 0. \qquad (2.37)$$

α_1 is an elastic constant of the material and is equal to

$$[(1 - 2\nu)/(2 - 2\nu)]^{\frac{1}{2}},$$

where ν is Poisson's ratio. The particle motion in Rayleigh waves is in the plane which is perpendicular to the surface along which the waves are travelling and parallel to the direction of propagation. For sinusoidal Rayleigh waves the trajectory of each particle is an ellipse.

When a dilatation wave is incident on a free surface two waves are generated on reflection, one of which is a dilatational wave reflected at an angle equal to the angle of incidence (α), while the other is a distortional wave which is reflected at a smaller angle (β), where $\sin\beta/\sin\alpha = c_2/c_1$. Similarly if a distortional wave is incident on a free boundary at an angle β both a distortional and a dilatational wave are in general reflected. The distortional wave is here reflected at an angle β and the dilatational wave at an angle α given by the above sine relation. A wave incident at the interface between two media will, in general, produce four waves; a dilatational and a distortional wave being propagated in each medium. The relations between the amplitudes of the incident, reflected, and refracted waves are given in equations (2.41) to (2.58) of Chapter II.

In Chapter III the types of wave which can be propagated along rods have been considered. There are three types, longitudinal, torsional, and flexural. If the wavelength is long compared with the lateral dimensions of the rod longitudinal and torsional waves travel with constant velocities. The velocity of propagation of longitudinal waves, c_0, is given by $(E/\rho)^{\frac{1}{2}}$, where E is Young's modulus, whilst the velocity of torsional waves in a rod of circular cross-section is $(\mu/\rho)^{\frac{1}{2}}$. The velocity of propagation of flexural waves depends on the wavelength and for very long wavelengths is given by $c = 2\pi c_0 K/\Lambda$, where Λ is the wavelength, $c_0 = (E/\rho)^{\frac{1}{2}}$, and K is the radius of gyration of a cross-section about the neutral axis of the rod.

When the wavelength becomes comparable with the lateral dimensions of the bar the velocity of longitudinal waves depends on the wavelength and at very short wavelengths they travel with the velocity of Rayleigh surface waves (cf. Figs. 14 and 15). The velocity of torsional waves is independent of the wavelength so long as the bar vibrates in its fundamental mode, i.e. so long as each section rotates as a whole about its centre. In practice it is found that it is only this fundamental mode that is generated. The velocity of flexural waves also approaches that of Rayleigh surface waves when the wavelength becomes small compared with the lateral dimensions of the bar (cf. Figs. 16 and 17).

The velocity of propagation of plane longitudinal waves in plates (c_L) is given by

$$c_L = \left[\frac{4\mu(\lambda+\mu)}{(\lambda+2\mu)\rho}\right]^{\frac{1}{2}} \tag{3.90}$$

$$= \left[\frac{E}{\rho(1-\nu^2)}\right]^{\frac{1}{2}} \tag{3.91}$$

when the wavelength is long compared with the thickness of the plate. For short wavelengths, these too travel with the velocity of Rayleigh surface waves (cf. Fig. 20).

CHAPTER IV

EXPERIMENTAL INVESTIGATIONS
WITH ELASTIC MATERIALS

ALTHOUGH much of the theory given in the previous chapter dates from the end of the last century, it is only comparatively recently that experimental techniques have become available for testing many of the results of this theory. Electronic methods have greatly facilitated both the production and the detection of high-frequency elastic waves, and this chapter will give a short survey of the experimental results on the propagation of stress waves through materials whose behaviour does not depart seriously from perfect elasticity. Experimental work concerned with departures from perfect elasticity will be described in Chapter VI.

Hopkinson pressure bar

In the pre-electronic era, experimental work on the propagation of elastic waves in solids was largely confined to the detection of seismic waves, and to the investigation of vibrations at audible frequencies in acoustical experiments. Bertram Hopkinson (1914) was among the first to investigate the propagation of stress pulses on a laboratory scale, and he did this in order to study the nature of the pressure-time relations when an explosive was detonated or when a projectile impinged on a hard surface. His apparatus, which has become known as the Hopkinson pressure bar, is an application of the simple theory of stress propagation of elastic pulses in a cylindrical bar where the length of the pulse is great compared with the radius of the bar. An electrical version of the Hopkinson bar devised by Davies (1948) will be described in the next section. This has enabled the nature of the propagation of pulses of lengths comparable with the lateral dimensions of the bar to be investigated experimentally.

Hopkinson's apparatus consisted of a cylindrical steel bar several feet in length and about an inch in diameter which was suspended in a horizontal position by four threads so that it could swing in a vertical plane. At one end of the bar a short cylindrical pellet known as the *time-piece* was wrung on, and the transient pressure was applied at the other end, known as the firing end of the bar. The time-piece was of the same diameter and of the same type of steel as the bar. One face of the time-piece and the end face of the bar to which it was attached were ground flat and the time-piece was held on by magnetic attraction or by smearing a little grease on the two ground faces.

When a bullet impinges on the firing end of such a bar, or when an explosive charge is detonated in contact with it, a compression pulse travels down the bar and, as was shown in the previous chapter, the pulse will travel without distortion, so long as the diameter of the bar is small compared with the length of the pulse, and the material of the bar is not stressed beyond its proportional elastic limit. Such a compression pulse will be transmitted through the joint between the bar and the time-piece without change in form, and will then be reflected at the free end of the time-piece as a pulse of tension. This reflected tension pulse will travel back through the tail of the incident compression pulse, and as soon as a tensile stress is built up across the joint between the bar and the time-piece, the latter will fly off with the momentum trapped in it. In Hopkinson's experiments this momentum was measured by capturing the time-piece in a ballistic pendulum, and at the same time the momentum remaining in the bar could be determined from the amplitude of swing of the bar.

Fig. 21 illustrates the stress distribution at various stages when a plane compression pulse of arbitrary shape is reflected at a free surface. The resultant stress at any point in the bar during reflection is obtained by adding the stresses due to the incident and reflected pulses, which are shown by the thin lines in the figures. The resultant stress is shown by the thickest line in each figure, whilst the broken line corresponds to the portion of the pulse which has already been reflected. (*a*) shows the pulse approaching the free boundary; (*b*) shows the distribution when a part of the pulse has been reflected but the stress still consists entirely of compression; (c) is a slightly later stage when some tension has been set up near the boundary; this tension has spread in (d), and in (e) half the pulse has been reflected and the stress is entirely in the form of tension. In (f) the reflection is completed and the tension pulse which emerges is of the same shape as the incident compression pulse.

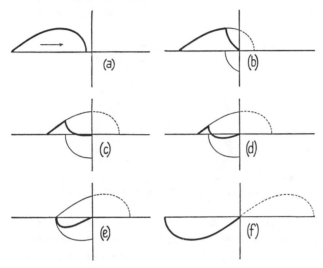

Fig. 21. Reflection of compression pulse at free boundary.

The reflection of a compression pulse at the free face of the time-piece will result in stress distributions similar to those shown in the figure, but when the time-piece is less than the length of the pulse, the time-piece will separate from the bar before the reflection is completed. When the time-piece leaves the bar, it has momentum trapped in it which corresponds to a portion of the pulse which is twice the length of the time-piece, and, as can be seen from (e) in Fig. 21, a time-piece half the length of the pulse will trap all the momentum, leaving the bar at rest. This gives a method of measuring the duration of the pulse, since this can be obtained if the minimum length of time-piece which will leave the bar undisturbed and the velocity of longitudinal waves in the material of the bar are known.

By measuring the momentum trapped in time-pieces of different lengths, the areas of the pressure-time curves for different intervals can be obtained. The precise shape of the pressure-time curve cannot, however, be deduced from such measurements since the points of commencement of the different intervals are not known. This is illustrated in Fig. 22, which shows three shapes of pulse which would be consistent with one set of observations. These curves are such that any horizontal

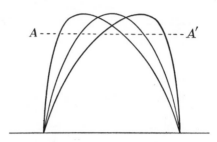

FIG. 22. Different shapes of pulse consistent with the same set of observations on the Hopkinson bar.

line such as AA' cuts off equal intercepts on each curve, and the length of these intercepts will correspond to twice the length of the time-piece. The momentum captured will be given by the area of the pressure-time curve between the ordinates which pass through the points of intersection, and their areas will be the same for all three curves. The maximum value of the pressure in the pulse can, however, be determined, since it is the limit of the average pressure for very short time-pieces. (In Fig. 22 this corresponds to the height of AA' when it becomes tangential to the three curves.)

The Hopkinson bar has been used by the Research Department, Woolwich (Robertson, 1921) for measuring the pressures set up by the detonation of various explosives, and Landon and Quinney (1923) describe some experiments with a Hopkinson bar, the firing-end of which was coned. When a sufficiently long time-piece was used with such a bar, it was found that the bar, instead of moving forward when the time-piece was thrown off, moved backwards. They point out that the reason for this phenomenon is that a compression pulse travelling along a conical bar will produce a tension tail, and when the region of the pulse where the change from compression to tension occurs reaches the joint the time-piece will fly off leaving the tail of the pulse, which has negative momentum, associated with it in the bar. The simple theory of propagation along a conical bar was discussed at the end of Chapter III.

Davies bar

Although the Hopkinson bar has the advantage of simplicity, it suffers from two serious limitations. Firstly, as shown in the previous section, it does not give the shape of the pressure-time curve of a pulse but only its duration and the value of the maximum pressure. Secondly, the tension necessary to break the joint between the bar and the time-piece introduces an unknown variable into the experiments and precludes the use of the apparatus for the measurement of pulses of small amplitude.

R. M. Davies (1948) has, however, devised a pressure bar in which the measurements are made electrically and this apparatus gives a continuous record of the longitudinal displacement produced by the pressure pulse at the free end of the bar. For a plane wave travelling down a bar it was shown in Chapter III that the longitudinal stress σ_{xx} is proportional to the particle velocity $\partial u/\partial t$, the relation being

$$\sigma_{xx} = \rho c_0 \frac{\partial u}{\partial t} \quad \text{(see equation (3.8))}.$$

Because of reflection, the free end of the bar has twice this particle velocity, and if the longitudinal displacement and particle velocity are denoted by ξ and $\dot{\xi}$ respectively we have

$$\dot{\xi} = 2 \frac{\partial u}{\partial t} = 2 \frac{\sigma_{xx}}{\rho c_0}. \tag{4.1}$$

With the Davies bar the (ξ, t) curve may be obtained directly, and by differentiating this curve the pressure-time curve for the pulse may be determined. If instead of measuring the longitudinal displacement of the end of the bar the radial displacement ζ' is measured at some point along the bar we have from the definition of Poisson's ratio ν

$$\zeta' = \frac{\nu a \sigma_{xx}}{E}, \qquad (4.2)$$

where a is the radius of the bar and E is Young's modulus; thus the pressure-time curve can be obtained from the (ζ', t) curve simply by multiplying the ordinates by $\nu a/E$.

As in the case of the Hopkinson bar the Davies apparatus will only give a true record of the pressure applied to it when (a) the stresses do not anywhere exceed the elastic limit of the steel and (b) the pressure does not change so suddenly that the wavelengths associated with the pressure pulse become comparable with the radius of the bar. Fig. 23 shows the general arrangement of the Davies bar. The longitudinal displacement ξ of the end surface of the bar is measured by using the bar as the earthed conductor of a parallel-plate condenser. The insulated conductor consists of a metal plate mounted in the 'bar condenser unit'. This unit slides freely on to the end of the bar and holds the insulated plate parallel to the end surface. For any slow movements of the bar the two conductors move together whilst on the arrival of a pressure pulse the end of the bar moves freely whilst the insulated plate remains instantaneously at rest as a result of its inertia. The insulated plate is charged up to a high voltage through the 'condenser feed unit'. This contains a condenser resistance arrangement of long time constant so that the charge on the insulated plate can only change very slowly, and any rapid change in the capacity of the parallel-plate condenser consequently results in corresponding changes in the potential difference across it. When the relative change in capacity is small this potential difference is directly proportional to the displacement of the end surface of the bar. These changes in potential difference are amplified by a wideband amplifier and fed on to a cathode-ray oscillograph where they are recorded photographically.

The cathode-ray oscillograph used is of the double beam type, there being two spots on the oscillograph screen which have identical horizontal motions but which can be given independent vertical deflexions. The amplified electrical output from the parallel-plate condenser unit is used to deflect one oscillograph spot whilst a radio-frequency timing wave is fed on to the other spot. The horizontal motion of the spots is

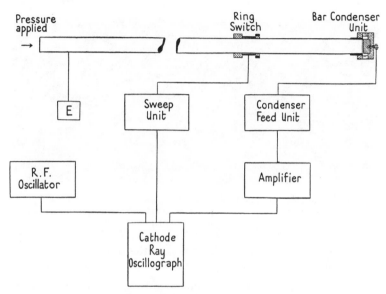

FIG. 23. General arrangement of the Davies pressure bar.

produced by a 'sweep unit' which is triggered by an inertia switch on the bar. This switch consists of an insulated metal ring which slides freely on the bar and makes contact with metal studs screwed into the bar. The arrival of the pressure pulse separates the ring from the studs and this fires a gas-filled thermionic valve in the sweep unit. In addition to producing a sweep of the oscillograph spots across the screen the sweep unit applies a transient positive voltage to the control grid of the cathode-ray tube so that the brightness of the spots is increased whilst they are traversing the screen.

Fig. 24 shows the type of record obtained with this apparatus. This photograph was obtained with a 1-inch diameter bar 6 feet in length when a no. 8 detonator was fired at one end of the bar.

The main pulse may be seen to have a duration of about 25 microseconds, and to be followed by a series of oscillations. The original pressure pulse produced by the detonator has here been considerably lengthened by the effects of dispersion in the bar. This dispersion also results in the lagging behind of high-

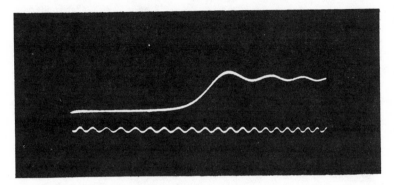

FIG. 24. Oscillogram obtained with Davies bar from a no. 8 detonator. Lower trace: timing wave of period 5·3 microseconds. (Oscillograph spots travelling from left to right.)

frequency components which appear as the series of oscillations seen at the end of the pulse.

The effects of this dispersion on pulses of different shapes and durations have been investigated theoretically by Davies (1948) and he has shown that for a bar 1 inch in diameter and 2 feet in length the distortion produced in a pulse originally 20 microseconds in length results in errors of the order of 2 or 3 per cent. in the pressure measurements. Davies confirmed these results experimentally by measurements of bullet impacts and detonation waves in gaseous mixtures where the pressures may be calculated theoretically.

Two other condenser units are described by Davies and these are of the cylindrical type. In both these condenser units an insulated cylindrical metal tube is held with its axis along the axis of the bar. The first type of unit is used at the end of the bar so that it measures longitudinal displacements, whilst the second type is placed some way down the bar and measures radial displacements. Davies has shown that for pulses of short duration both types of cylindrical unit accentuate the effects of distortion more than does the parallel-plate unit. These units are thus only useful with long pulses. The cylindrical unit measuring longitudinal displacement has the advantage, however, of constant sensitivity even when the displacements are large, whilst the unit measuring radial displacement gives an output proportional to the pressure (see equation (4.2)) and thus obviates the necessity for differentiating the displacement-time curve.

Torsional pressure bar

It was shown in the previous chapter that the velocity of propagation of harmonic torsional waves in a circular cylinder is independent of their wavelength so long as the motion takes place in the fundamental mode of vibration of the bar, i.e. so long as the angular displacement is a linear function of the distance from the axis of the bar. A torsional pulse of this type should therefore be propagated down a bar without change in form, and a pressure bar in which the applied pressure is made to excite torsional waves instead of longitudinal ones might therefore be expected to give more faithful measurements of pressures which are changing rapidly.

Such a pressure bar has been designed by R. M. Davies and J. D. Owen (Owen and Davies, 1949; Davies and Owen, 1950; Owen, 1950). With this apparatus the angular displacement of a small optically polished flat on the surface of a cylindrical steel bar is recorded photographically, a high-speed revolving mirror being used to give a time sweep. These workers have found that for the torsional pulse excited by the impact of a bullet on the side of the bar the motion takes place in the fundamental mode and the pulse is propagated down the bar without change in form. They have shown that this apparatus will record large changes in pressure which take place in less than a microsecond.

Ultrasonic measurements

The experimental results on the propagation of stress pulses along cylindrical bars were shown by Davies (1948) to be in agreement with the theoretical predictions of the Pochhammer–Chree theory. Measurements of the phase velocity of ultrasonic harmonic waves give a more direct verification of the theory, and during recent years a number of workers have carried out such measurements. The method used in all these investigations was to set a rod of the material into resonance; the phase velocity was then obtained as the product of the frequency and the wavelength. With a given rod a number of resonances could be observed corresponding to the fundamental frequency and a series of harmonics.

Giebe and Scheibe (1931) set quartz rods in vibration by utilizing their piezo-electric properties. The resonant frequencies were observed by having the rods in a rarefied atmosphere with a small gap between the crystal and one of the energizing electrodes. At resonance a glow discharge was found to occur in this gap. Giebe and Blechschmidt (1933) used the magneto-strictive effect to excite rods and tubes of nickel and nickel-iron alloy. In their apparatus two coils were placed round the rod and a high-frequency alternating current was passed through one of the coils whilst the other coil was used as a detector, whose output was rectified and then measured with a galvano-meter.

Röhrich (1932), Schoeneck (1934), and Shear and Focke (1940) used quartz crystals to excite rods of other materials and measured the wavelengths at resonance directly from the standing wave patterns obtained when lycopodium powder was sprinkled on the rods. Röhrich worked with cylinders of poly-crystalline steel, copper, aluminium, and brass, and also with cylinders of glass. Schoeneck used single crystal specimens of zinc, cadmium, and tin and also specimens of polycrystalline zinc, whilst Shear and Focke made their measurements with cylinders of polycrystalline silver, nickel, and magnesium.

All these investigators found that when the wavelength was several times the diameter of the bar the experimental results were in good agreement with the values predicted by the Rayleigh correction (see equation (3.60) and curve 1A in Fig. 14). At higher frequencies, however, when the wavelengths become of the same order as the diameter of the cylinder, the observed phase velocities were found to be lower than those obtained on applying the Rayleigh correction.

Giebe and Blechschmidt (1933) gave a theory of the propagation of longitudinal waves along a cylinder, and the phase

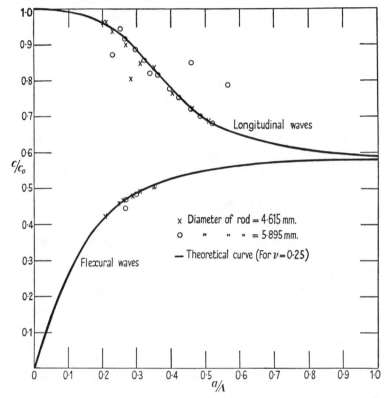

FIG. 25. Experimental results for magnesium rods compared with theoretical curves.

velocities produced by this theory were in better agreement with the observed values. Röhrich (1932), however, found that this theory too became inadequate at the highest frequencies. The Giebe and Blechschmidt treatment, and the reasons why it is only approximate, have been discussed in Chapter III. Shear and Focke (1940) made their measurements in order to test the

Giebe and Blechschmidt treatment, and they also found that it failed at the higher frequencies. The work of Bancroft (1941) and Hudson (1943) (loc. cit.) has finally produced exact values for the velocities of longitudinal and transverse waves in cylinders in terms of the Pochhammer–Chree equations. The experimental results of Shear and Focke were found to be in good agreement with these predicted velocities.

Fig. 25 shows a comparison between the theoretical curves and the velocities observed by Shear and Focke for two magnesium rods of different diameters; the value of Poisson's ratio ν has been taken as 0·25 for the theoretical curves. The results are plotted in non-dimensional form, the ratio c/c_0 being given for different values of the ratio a/Λ. (c is here the phase velocity of waves of wavelength Λ, c_0 is the velocity of longitudinal waves of infinite wavelength, and a is the radius of the bar; cf. Fig. 16). It may be seen that the agreement is very good with the exception of a few isolated points which presumably correspond to other modes of vibration. One of the main difficulties in the experimental work was that in exciting the cylinders flexural, torsional, and longitudinal waves were generally produced simultaneously and the wave patterns observed tended to be very complex.

Stanford (1950) has carried out measurements of the longitudinal vibrations of cylindrical aluminium bars of diameters between $\frac{1}{2}$ inch and 12 inches. The bars were clamped at their midpoints and excited to resonance by means of thin nickel tubes which were driven magnetostrictively. The frequencies used were between 5 and 50 kilocycles per second, and Stanford found that the curve of c_p plotted against the radius of the bar followed the theoretical curve 1 shown in Fig. 14 very closely. At the higher frequencies he obtained some evidence of resonances corresponding to the higher modes shown by curves 2 and 3.

PART II

STRESS WAVES IN IMPERFECTLY ELASTIC MEDIA

CHAPTER V

INTERNAL FRICTION

THE theory described in the first part of this monograph is based on Hooke's law. The equations of motion of a solid were obtained by equating the products of the masses and accelerations to the elastic forces, and it was assumed that no other forces came into play. Many solids do not depart seriously from perfectly elastic behaviour for small deformations and, as was shown in Chapter IV, the observed results often agree well with the predictions of the elastic theory.

When materials are set in vibration some of the elastic energy is, however, always converted into heat, and the various mechanisms by which this takes place are collectively termed internal friction. Thus when a solid specimen vibrates, its free oscillations decay even when it is isolated from its environment. The amplitude of vibration of a specimen should, in the absence of internal friction, increase indefinitely when driven at its resonant frequency by an external alternating force. In practice the amplitude is always found to assume a finite value.

For liquids and gases the dissipative forces are due to viscosity and thermal conduction, and these effects can be treated analytically. In solids the behaviour is found to be much more complex and to vary considerably with the nature of the solid. There is at present no satisfactory theory of internal friction in solids, and more experimental data are required.

Definitions

The most direct method of defining internal friction is as the ratio $\Delta W/W$, where ΔW is the energy dissipated in taking a specimen through a stress cycle and W is the elastic energy stored in the specimen when the strain is a maximum. This ratio is sometimes called the 'specific damping capacity' or the 'specific loss' and can be measured for a stress cycle without any assumptions being made about the nature of internal friction. The value obtained is, however, generally found to depend on the amplitude and the speed of the cycle, and often also on the past history of the specimen.

There are several indirect methods of defining internal friction and these depend on the assumption that the restoring forces are proportional to the amplitude of vibration, whilst the dissipative forces are proportional to the velocity. When these conditions apply, the ratio between successive free oscillations is constant and the natural logarithm of this ratio, Δ', which is called the logarithmic decrement, is taken as a measure of the internal friction. As shown below, if Δ' is taken as the natural logarithm of the ratio of successive excursions on the same side of the equilibrium position it is equal to half the specific loss when the damping is low.

Another indirect measure of internal friction is given by the sharpness of the resonance curve under forced vibration. If a specimen is driven by a sinusoidal force of fixed amplitude, the frequency of which can be varied, and the amplitude of vibration of the specimen is plotted against frequency the curve shows a maximum at the resonant frequency N and falls off on either side of it. The lower the internal friction of the specimen the sharper this resonance peak is found to be, and if ΔN is the change in the impressed frequency necessary to change the amplitude from half its maximum value at one side of the resonant frequency to half its maximum value on the other side, $\Delta N/N$ is a measure of the internal friction. For a linear system with low damping $\Delta N/N$ is equal to $\sqrt{3}/2\pi$ times the specific loss.

We shall now show how the relations between these various definitions of internal friction are obtained for a specimen in which the elastic restoring force is proportional to the displacement and the dissipation is proportional to the velocity. The equation of motion of such a specimen may be written

$$P = M_1\ddot{\xi} + \eta_1\dot{\xi} + E_1\xi, \tag{5.1}$$

where P is the impressed force and ξ is the displacement. The first term of the right-hand side of the equation is an inertia term, M_1 depending on the mass and shape of the specimen; the second term is a damping term, η_1 involving the ratio between the velocity and the damping force; the third term is an elastic term, E_1 depending on the elastic modulus and shape of the specimen.

For free oscillations P is zero and the general solution of (5.1) is then

$$\xi = A \exp\left\{[-\eta_1 + \sqrt{(\eta_1^2 - 4E_1 M_1)}]\frac{t}{2M_1}\right\} + $$
$$+ B \exp\left\{[-\eta_1 - \sqrt{(\eta_1^2 - 4E_1 M_1)}]\frac{t}{2M_1}\right\}, \quad (5.2)$$

where A and B are complex constants which depend on the initial conditions of the motion. The nature of the motion will depend on whether η_1^2 is greater or less than $4E_1 M_1$. If it is greater the motion will not be oscillatory; if it is less we have on taking the real parts of (5.2)

$$\xi = C \exp\left(\frac{-\eta_1 t}{2M_1}\right)\cos(p_1 t + \beta), \quad (5.3)$$

where

$$p_1^2 = \frac{E_1}{M_1} - \frac{\eta_1^2}{4M_1^2} \quad (5.4)$$

and β and C are functions of A and B.

The frequency of the oscillation is equal to $p_1/2\pi$ and from (5.4) it may be seen that the frequency as well as the amplitude of the vibration will depend on η_1. The amplitude is reduced by a factor $\exp(\eta_1 t/2M_1)$ in time t, so that in one period of the oscillation it will be reduced by a factor $\exp(\eta_1 \pi/M_1 p_1)$. The logarithmic decrement is thus given by

$$\Delta' = \frac{\eta_1 \pi}{M_1 p_1}. \quad (5.5)$$

When the damping is sufficiently small for η_1^2 to be neglected compared with $4E_1 M_1$, equation (5.4) shows that p_1^2 approaches E_1/M_1 and (5.5) then becomes

$$\Delta' = \frac{p_1 \pi \eta_1}{E_1}. \quad (5.6)$$

The elastic energy stored in the specimen is proportional to the square of the amplitude so that if ξ_1 and ξ_2 are successive amplitudes on the same side of the equilibrium position the specific loss $\Delta W/W$ will be given by $(\xi_1^2-\xi_2^2)/\xi_1^2$. When this is small compared with unity we have

$$\frac{\Delta W}{W} = \frac{\xi_1^2-\xi_2^2}{\xi_1^2} \simeq \frac{2(\xi_1-\xi_2)}{\xi_2} \simeq 2\ln\left(\frac{\xi_1}{\xi_2}\right) = 2\Delta'. \qquad (5.7)$$

Thus in free oscillations, if the damping is low, the specific loss is equal to twice the logarithmic decrement.

We shall now consider forced oscillations where P in equation (5.1) is a force varying sinusoidally with time with a frequency of $p/2\pi$ so that

$$P = P_0 \sin pt. \qquad (5.8)$$

The solution of (5.1) is now the sum of two terms, a complementary function and a particular integral. The complementary function is given by equation (5.3) and corresponds to transient conditions; as t increases this function approaches zero. The particular integral which corresponds to the steady state solution is

$$\xi = \frac{P_0}{pZ}\sin(pt-\delta), \qquad (5.9)$$

where

$$Z^2 = \left(\frac{E_1}{p} - M_1 p\right)^2 + \eta_1^2 \qquad (5.10)$$

and

$$\tan\delta = \frac{p\eta_1}{(E_1 - M_1 p^2)}. \qquad (5.11)$$

The above solution is analogous to that obtained for an electrical circuit with inductance, capacitance, and resistance; M_1, $1/E_1$, and η_1 correspond respectively to these three electrical quantities, where the displacement ξ corresponds to the charge. Z then corresponds to the electrical impedance of the circuit and is called the mechanical impedance of the vibrating system. From equation (5.9) it may be seen that the amplitude is a maximum when pZ is a minimum and this corresponds to

$$p^2 = \frac{E_1}{M_1} - \frac{\eta_1^2}{2M_1^2}. \qquad (5.12)$$

The applied frequency at which the amplitude is a maximum is thus not equal to p_1 the natural frequency of the vibratory system (see equation (5.4)). To obtain the 'half-breadth' of the resonance peak we must find the two values of p for which the amplitude of ξ has half its maximum value. From (5.9), (5.12), and (5.4) it may be shown that this maximum value is $P_0/\eta_1 p_1$ so that the half-values correspond to

$$(E_1 - M_1 p^2)^2 + \eta_1^2 p^2 = 4\eta_1^2 p_1^2.$$

The two solutions of this equation become on substituting from (5.4)

$$p^2 = \left(\frac{E_1}{M_1} - \frac{\eta_1^2}{2M_1^2}\right) \pm \frac{\sqrt{3}\,\eta_1 p_1}{M_1}$$

so that if the two values of p are p_2 and p_3, we have

$$p_2^2 - p_3^2 = \frac{2\sqrt{3}\,\eta_1 p_1}{M_1}. \tag{5.13}$$

The half-breadth of the resonance peak is given by $\Delta N/N$, where N is the resonant frequency and ΔN is the difference between the two frequencies corresponding to p_2 and p_3. Now p in each case is 2π times the frequency so that if the damping is low we have

$$\frac{\Delta N}{N} \backsimeq \frac{p_2 - p_3}{p_1} \backsimeq \frac{p_2^2 - p_3^2}{2p_1^2},$$

and from (5.13) and (5.5) we then have

$$\frac{\Delta N}{N} \backsimeq \frac{p_2^2 - p_3^2}{2p_1^2} = \frac{\sqrt{3}\,\eta_1}{M_1 p_1} = \frac{\sqrt{3}\,\Delta'}{\pi}. \tag{5.14}$$

Thus the half-breadth of the resonance peak equals $\sqrt{3}/\pi$ times the logarithmic decrement Δ', or $\sqrt{3}/2\pi$ times the specific loss (see equation (5.7)). By analogy with the electrical case a quantity Q is sometimes used as a measure of the sharpness of resonance: this is defined as

$$\frac{1}{Q} = \frac{1}{\sqrt{3}} \frac{\Delta N}{N};$$

thus

$$Q = \frac{\pi}{\Delta'} = \frac{2\pi}{\Delta W/W}. \tag{5.15}$$

The work done by the impressed force in taking the specimen through a sinusoidal stress cycle of period τ is given by

$$\Delta W = \int_0^\tau P \frac{d\xi}{dt} dt.$$

From (5.8) and (5.9) this becomes

$$\Delta W = \frac{P_0^2}{Z} \int_0^\tau \sin pt \cos(pt-\delta)\, dt$$

$$= \frac{P_0^2 \pi}{Zp} \sin \delta \quad \left(\text{since } \tau = \frac{2\pi}{p}\right). \qquad (5.16)$$

From (5.11) and (5.10) it may be seen that $\sin \delta = \eta_1/Z$, so that

$$\Delta W = \frac{\pi P_0^2 \eta_1}{Z^2 p}. \qquad (5.17)$$

The elastic energy stored in the specimen when the displacement is a maximum is given by $W = \frac{1}{2}E_1(\xi_{\max})^2$ and from (5.9) we therefore have

$$\frac{\Delta W}{W} = \frac{2\pi\eta_1 p}{E_1} \qquad (5.18)$$

or

$$\Delta' = \frac{\pi\eta_1 p}{E_1}, \qquad (5.19)$$

which is similar to equation (5.6) for free oscillations.

Methods of measurement

The definitions of internal friction discussed in the previous section suggest various ways in which the internal friction of a specimen can be measured. Thus the specific loss can be determined directly as the heat produced when a specimen is taken round a stress cycle. This has been done for steel by Hopkinson and Williams (1912) and more recently by Föppl (1936) who measured the difference in temperature between the middle and ends of a test-piece when undergoing cyclic deformation. This difference in temperature is proportional to the rate at which heat is being generated in the specimen and conducted away. In order to obtain absolute values, the apparatus was calibrated by passing an electric current through the specimen when at rest, and observing the temperature difference for a known heat dissipation. Hopkinson and Williams used stress ranges of up to 30 tons per square inch and frequencies up to 120 cycles per second. From the peak value of the stress the maximum elastic energy stored in the specimen could be calculated and the specific loss $\Delta W/W$ could hence be determined. The main disadvantage of this method is that large forces are required to obtain measurable temperature differences and the apparatus must be designed on an engineering scale.

Indirect measurements of internal friction can be made by observing the logarithmic decrement of a specimen oscillating freely or the sharpness of resonance under forced vibrations. A large number of measurements have been made by these two methods and these will be discussed in Chapter VI. Another method of investigating internal friction which is more closely related to the general subject of this monograph is the measurement of the attenuation of a stress wave when it travels through a solid.

For a plane sinusoidal wave of small amplitude the attenuation is found to be exponential, so that if the pressure amplitude is P_0 at the origin it will be $P_0 \exp(-\alpha x)$ after the wave has travelled a distance x. α is here the attenuation constant and is a measure of the internal friction of the material. The energy flux for a plane wave of pressure amplitude P is given by $P^2/2\rho c$, where ρ is the density of the material and c is the velocity of propagation. If we consider a slab of material of unit cross-section normal to the wave and of thickness δx the energy entering per unit time is

$$\frac{P_0^2 \exp(-2\alpha x)}{2\rho c}$$

and the energy leaving per unit time

$$\frac{P_0^2 \exp(-2\alpha(x+\delta x))}{2\rho c}.$$

Thus the energy dissipated per second in the slab is approximately

$$\frac{P_0^2 \alpha \, \delta x}{\rho c} \exp(-2\alpha x),$$

and ΔW the energy dissipated during one cycle is therefore given by

$$\Delta W = \frac{2\pi P_0^2 \alpha \delta x}{\rho c p} \exp(-2\alpha x), \qquad (5.20)$$

where the frequency is $p/2\pi$.

The energy density in the medium is $P^2/2\rho c^2$, so that W, the maximum elastic energy stored in the slab, is given by

$$W = \frac{P_0^2 \delta x}{2\rho c^2} \exp(-2\alpha x); \qquad (5.21)$$

from (5.20) and (5.21) we therefore have

$$\alpha = \frac{p}{4\pi c} \frac{\Delta W}{W}, \qquad (5.22)$$

so that α may be related to the specific loss and hence to the other quantities used for defining internal friction. The wave propagation method has been used in recent years by several workers for measuring internal friction with materials which can be obtained in the form of strips or wires, and the results of these measurements will be described in the next chapter.

The behaviour of 'visco-elastic' solids

In discussing equation (5.1) for a vibrating body, it has been assumed that the elastic restoring force is proportional to the displacement and the dissipative force is proportional to the velocity. It was pointed out that E_1 in equation (5.1) depends on the elastic constants whilst η_1 depends on the dissipative forces, both quantities also involving the dimensions of the specimen. The nature of these dissipative forces was, however, not discussed and we shall now consider how the internal friction depends on frequency if we assume that these forces are of a purely viscous nature. In treating this problem we must first decide on the parameters on which these viscous forces will depend. Maxwell (1890), in discussing the nature of viscosity in gases, suggested that whilst the relation between the stress σ and the strain ϵ in an elastic solid was $\sigma = E^*\epsilon$, where E^* is the appropriate elastic constant, the relation for a real solid is more closely predicted by

$$\frac{d\sigma}{dt} = E^* \frac{d\epsilon}{dt} - \frac{\sigma}{\tau}, \tag{5.23}$$

where τ is the 'relaxation time' of the solid.

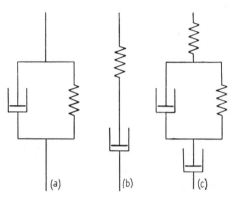

FIG. 26. Models of visco-elastic solids. (a) Voigt solid;
(b) Maxwell solid; (c) more general solid.

Thus when a force is applied for times short compared with τ, the material behaves like an elastic solid, whilst for times long compared with τ the behaviour is like that of a viscous liquid of viscosity $E^*\tau$. A solid which obeys (5.23) is referred to as a 'Maxwell solid' and may be represented by a model consisting of a spring in series with a viscous element often called a 'dashpot' (see Fig. 26 (b)). The spring is assumed to obey Hooke's law, whilst the dashpot can be considered as a piston being drawn through a liquid which obeys Newton's law of viscosity so that the velocity is proportional to the applied force.

We shall return to the behaviour of Maxwell solids, but shall first discuss the other type of coupling between elastic and viscous elements which was originally considered by Meyer (1874) and later extended by Voigt (1892). Voigt assumed that the stress components in a solid could be expressed as the sum of two sets of terms, the first set being proportional to the strains and the second set proportional to the rate of change of the strains. Thus in the equations for an aeolotropic solid (equations (2.2), Chapter II) each stress component would be the sum of twelve terms, there being six terms of the form $c_{rs}\epsilon_{ij}$ and six terms of the form $a_{rs}(\partial/\partial t)\epsilon_{ij}$; the c-coefficients are the elastic constants of the material and the a-coefficients the corresponding viscous constants. For an isotropic material the latter reduce to two viscous constants which correspond to Lamé's constants; these will be denoted by λ' and μ'. The elastic equations (2.3) then become

$$\left.\begin{aligned} \sigma_{xx} &= \lambda\Delta + 2\mu\epsilon_{xx} + \lambda'\frac{\partial\Delta}{\partial t} + 2\mu'\frac{\partial\epsilon_{xx}}{\partial t}, \quad \text{etc.} \\ \text{and} \qquad \sigma_{yz} &= \mu\epsilon_{yz} + \mu'\frac{\partial\epsilon_{yz}}{\partial t}, \quad \text{etc.} \end{aligned}\right\} \qquad (5.24)$$

These equations lead to relations similar to those obtained for an elastic solid, but the operator $\lambda + \lambda'(\partial/\partial t)$ takes the place of λ, and $\mu + \mu'(\partial/\partial t)$ takes the place of μ; in most cases this leads to rather complicated equations.

For a solid of this type, which is known as a Voigt solid, the stress is the sum of two terms, one proportional to the strain and the other proportional to the rate of change of strain. This behaviour can be described by a model of the type shown in Fig. 26 (a) with a spring and dashpot in parallel. In torsional deformation only shear is involved, and the 'stiffness' of the spring will be given by the shear modulus μ and the viscosity due to the dashpot will be given by μ'.

Thus if two opposing couples each of magnitude C act at opposite ends of a cylinder of length l and radius r and the relative displacement between the two end faces is θ, we have for a perfectly elastic solid

$$C = \tfrac{1}{2}\pi\mu r^4\theta/l \quad \text{(cf. equation (3.15))}, \qquad (5.25)$$

and for a Voigt solid

$$C = \frac{\tfrac{1}{2}\pi r^4}{l}\left(\mu\theta + \mu'\frac{d\theta}{dt}\right). \qquad (5.26)$$

If for simplicity we consider a disk of moment of inertia I suspended from a wire of length l and radius r and assume that the moment of inertia of the disk is large compared with that of the wire the equation of motion for torsional vibrations is:

$$I\ddot{\theta} + \frac{\pi r^4}{2l}(\mu'\dot{\theta} + \mu\theta) = 0. \qquad (5.27)$$

This is identical with equation (5.1) with $P = 0$ if ξ is taken as the angular displacement and

$$M_1 = I, \qquad E_1 = \frac{\pi r^4 \mu}{2l}, \qquad \eta_1 = \frac{\pi r^4 \mu'}{2l}.$$

Thus if the damping is small the logarithmic decrement of the system Δ' is given from equation (5.6) by

$$\Delta' = p_1 \pi \frac{\eta_1}{E_1} = p_1 \pi \frac{\mu'}{\mu}. \qquad (5.28)$$

Thus if the wire is behaving like a Voigt solid the logarithmic decrement is proportional to the frequency and also to the ratio μ'/μ.

In general, deformations involve dilatation as well as shear so that the four constants λ, λ', μ, and μ' must be used, and in order to treat these problems, simplifying assumptions have often been made about relations between them. Thus instead of using both Lamé's constants λ and μ, the compressibility k $(= \lambda + \frac{2}{3}\mu)$ and μ are used and the associated viscous constants are $(\lambda' + \frac{2}{3}\mu')$ and μ'. $(\lambda' + \frac{2}{3}\mu')$ is often denoted by χ and is known as the *dilatational viscosity* of the solid, μ' being the shear viscosity. It is then assumed that χ is negligible so that there is only a single viscous constant μ'. In the case of some substances, notably rubber-like materials, there seems to be some justification for this hypothesis, in that the effect of dilatational viscosity is small compared with that of shear viscosity. In general, however, the assumption does not appear to be true.

Another simplifying assumption which has sometimes been made (e.g. Quimby, 1925) is that $\lambda/\lambda' = \mu/\mu'$. If this were true there would again be only one independent viscous constant. There appears, however, to be little justification for making this assumption except by the general argument that the viscous processes might be expected to follow the elastic ones (see Thompson, 1933).

For uniaxial extension of a rod when the lateral surfaces are free from stress the ratio between stress and strain is given by Young's modulus $E = \mu(3\lambda+2\mu)/(\lambda+\mu)$ (see equation (2.4)). Thompson (1933) has shown that for a Voigt solid the relation between stress and strain will be

$$\sigma_{xx} = E\epsilon_{xx} + E'\frac{\partial\epsilon_{xx}}{\partial t}, \qquad (5.29)$$

where E' is given approximately by

$$\frac{E'}{E} = \frac{\lambda\mu\tau_1+(3\lambda^2+4\lambda\mu+2\mu^2)\tau_2}{(\lambda+\mu)(3\lambda+2\mu)}, \qquad (5.30)$$

$$\tau_1 = \frac{\lambda'}{\lambda} \quad \text{and} \quad \tau_2 = \frac{\mu'}{\mu}.$$

E' has been called the 'coefficient of normal viscosity' by Honda and Konno (1921). (If $\lambda'/\lambda = \mu'/\mu = \tau'$, (5.30) gives $E' = E\tau'$.)

The longitudinal vibrations of a rod which behaves like a Voigt solid will be represented by equation (5.1) if E_1 is equal to BE and $\eta_1 = BE'$, B being a quantity which depends on the shape of the rod and has the dimensions of length squared. Thus from (5.6) the logarithmic decrement Δ' will be given approximately by

$$\Delta' = \frac{p_1\pi E'}{E}, \qquad (5.31)$$

where p_1 is 2π times the frequency of oscillation.

We shall now return to the Maxwell solid, the stress-strain equation of which is (5.23); the corresponding mechanical model is shown in Fig. 26 (b). For simplicity we shall consider the longitudinal vibrations of a large mass M suspended from a rod of length l and cross-sectional area A and assume that the inertia of the rod is negligible compared with M. The equation of motion of the mass is then

$$M\ddot{\xi} + A\sigma = 0, \qquad (5.32)$$

where σ is the stress in the rod and ξ is the displacement of the mass. The strain in the rod, ϵ, will then be ξ/l so that equation (5.23) becomes

$$\frac{d\sigma}{dt} = \frac{E^*}{l}\frac{d\xi}{dt} - \frac{\sigma}{\tau}; \qquad (5.33)$$

eliminating σ and $d\sigma/dt$ from (5.32) and (5.33) we obtain

$$\frac{M}{A}\frac{d^3\xi}{dt^3}+\frac{M}{A\tau}\frac{d^2\xi}{dt^2}+\frac{E^*}{l}\frac{d\xi}{dt}=0. \tag{5.34}$$

We may integrate this so that

$$\frac{d^2\xi}{dt^2}+\frac{1}{\tau}\frac{d\xi}{dt}+\frac{E^*A}{Ml}\xi=\text{constant}. \tag{5.35}$$

(5.35) is similar in form to (5.1) and if the mass M oscillates it will perform damped sinusoidal oscillations which, by comparison with (5.3), can be seen to be given by

$$\xi=C\exp\left(-\frac{t}{2\tau}\right)\cos(p_1 t+\beta)$$

(where $p_1^2=E^*A/Ml-1/4\tau^2$). The logarithmic decrement Δ' is the natural logarithm of the ratio of two successive amplitudes on the same side of the equilibrium position. Since the amplitude is reduced by a factor $\exp(1/2\tau)$ per unit time, it will be reduced by $\exp(\pi/\tau p_1)$ in one oscillation, hence

$$\Delta'=\frac{\pi}{\tau p_1}. \tag{5.36}$$

As mentioned earlier, for a Maxwell solid the time of relaxation τ can be regarded as the ratio between its 'effective viscosity' η^* and its elastic modulus E^*, so that (5.36) may be written

$$\Delta'=\frac{\pi E^*}{\eta^* p_1}. \tag{5.37}$$

Thus for a Maxwell solid the logarithmic decrement varies *inversely* both with the frequency and with the effective viscosity of the material.

If we compare (5.37) with (5.28) we see that Maxwell solids and Voigt solids behave in opposite ways, Δ' being inversely proportional to frequency for Maxwell solids and directly proportional to frequency for Voigt solids. This furnishes a convenient test of whether real materials behave like one or other of these idealized models. As will be shown in the next chapter, for most solids Δ' does not behave in either manner and is often roughly independent of frequency. Thus although these models are useful in describing in a qualitative way how internal friction may arise from processes of a viscous nature, they are far too simplified to be quantitatively useful. Fig. 26 (c) shows a slightly more complicated arrangement which combines the features of the Voigt and Maxwell models. This model gives results more in accordance with the behaviour of real solids, but here again the quantitative agreement is found to be very poor except over small ranges of frequency.

To allow for the fact that a number of different relaxation phenomena may be taking place simultaneously in a solid, more complicated models have been considered. These consist of a number of Maxwell models joined in parallel or a number of Voigt models joined in series. The solid is thus considered as having a number of different relaxation times, or in the limit a continuous 'spectrum' of relaxation times. This treatment is mathematically equivalent to Boltzmann's approach which is discussed below.

The superposition principle

If a Maxwell solid is strained by an amount ϵ and held at this strain, the stress will relax with time. From equation (5.23) it may be seen that the stress decays exponentially, its value at time t being given by $E^*\epsilon \exp(-t/\tau)$. Boltzmann (1876) generalized this relation to materials for which the decay of stress was not necessarily exponential He suggested that the mechanical behaviour of a solid was a function of its entire previous history, and assumed that when a specimen had undergone a series of deformations the effect of each deformation was independent of that of the others, so that the resultant behaviour could be calculated by a simple addition of the effects that would occur if the deformations took place singly. This assumption has become known as the *Principle of Superposition*. Boltzmann assumed that the shear and dilatation may relax in different ways so that for deformations, such as uniaxial extension, in which both are involved the treatment becomes rather complicated. If, however, the deformation is in the form of torsion, which involves only shear, or the solid is one in which dilatational relaxation effects are small, the analysis is simpler.

It is assumed that if a specimen is strained by an amount $\epsilon(T)$ for a time δT at time T there will be a residual stress

$$-\epsilon(T)f(t-T)\,\delta T$$

at time t $(t > T)$ resulting from this. $f(t-T)$ is called a *memory function*, and is characteristic of the material and of the type of stress. The total stress σ is assumed to be the sum of the elements of stress remaining from the entire past history of the specimen together with the stress corresponding to the instantaneous value of the strain ϵ at time t. This contribution due to the value of ϵ is usually taken as proportional to ϵ, but in general is written as $F(\epsilon)$ so that the relation between the stress and strain becomes

$$\sigma(\epsilon, t) = F(\epsilon) - \int_{-\infty}^{t} \epsilon(T) f(t-T)\, dT. \tag{5.38}$$

Equation (5.38) is one form of the superposition principle in which the problem has been considered in terms of relaxation of *stress*, the stress at any time being given in terms of the strain history of the specimen. The problem can, however, be treated the other way round so that stress is the independent variable and the strain produced by a known stress history is then derived (see, for example, Leaderman, 1943). It has been shown (Gross, 1947) that the two treatments are mathematically equivalent, and it is generally a matter of convenience which of them is followed.

For a Maxwell solid $F(\epsilon)$ is given by $E^*\epsilon$ whilst the memory function $f(t-T)$ is $(E^*/\tau)\exp[(T-t)/\tau]$. Equation (5.38) then becomes

$$\sigma = E^*\epsilon - \frac{E^*}{\tau} \int_{-\infty}^{t} \epsilon(T)\exp\left(\frac{T-t}{\tau}\right) dT. \tag{5.39}$$

Thus if a previously undeformed specimen is strained an amount ϵ at time t_1, and the strain is held at this value the stress at time t is given by

$$\sigma = E^*\epsilon\left[1 - \frac{1}{\tau} \int_{t_1}^{t} \exp\left[\frac{T-t}{\tau}\right] dT\right] = E^*\epsilon \exp\left[\frac{t_1-t}{\tau}\right]$$

which is identical with the result obtained directly from equation (5.23).

For most real solids $f(t-T)$ cannot be expressed as a single exponential and better agreement with the experimental results is obtained by taking it to be the sum of a series of exponentials each of which has a different value of τ. This is equivalent to considering a mechanical arrangement made up of a series of Maxwell models joined in parallel. In the limit the memory function may be expressed as the integral of the form

$$f(t-T) = \int_0^\infty A(\tau)\exp\left(\frac{T-t}{\tau}\right) d\tau. \tag{5.40}$$

$A(\tau)$ is here a function of τ, $A(\tau)\,d\tau$ representing the 'number' of relaxation times between τ and $\tau+d\tau$, and the elastic modulus associated with them. The curve obtained by plotting $A(\tau)$ against τ gives the *relaxation spectrum* of the material, and in general this can be more easily related to the microscopic processes which produce relaxation than the memory function itself. Gross (1947) discusses the mathematical transformations required to determine the function $A(\tau)$ when $f(t-T)$ is known.

It should be noted that in the Maxwell model, when the strain is kept constant, the stress asymptotically approaches zero. This will apply to any number of Maxwell elements joined in parallel. For real solids it is found that the stress generally tends to a finite value, and this can be allowed for by having a second spring across the Maxwell element (see Fig. 27 (a)). In the case of the Boltzmann equation (5.33) this means that if ϵ is kept constant the integral approaches asymptotically a value $A\epsilon$ which is less than $F(\epsilon)$. The stress-strain relation for strains which have been maintained for a long time then becomes

$$\sigma = F(\epsilon)-A\epsilon,$$

this being the relation between σ and ϵ for the auxiliary spring.

The Voigt model does not show stress relaxation, since if the strain is fixed the stress is also fixed; if, however, a second spring is placed in series with it (Fig. 27 (b)) it becomes equivalent to a Maxwell model with a spring in series. Thus in Fig. 27 (a) let the stiffness of the spring in series with the dashpot be E_m (so that the extension produced by a tension P_1 in it is P_1/E_m), the stiffness of the auxiliary spring be E_a, and the 'viscosity' of the dashpot be η_m (so that if a tension P_1 is applied across it, it extends at the rate P_1/η_m). To obtain the relation between applied force P and the strain ϵ for this model we must add the forces in the

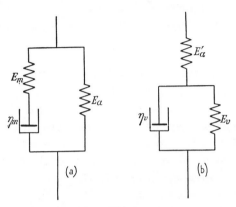

Fig. 27. Two equivalent mechanical models. (a) Auxiliary spring in parallel with Maxwell model. (b) Auxiliary spring in series with Voigt model.

two parallel arms. If we call the tension in the arm containing the Maxwell 'element' P_1, and that in the auxiliary spring P_2 we have

$$E_m P_1 + \eta_m \frac{dP_1}{dt} = E_m \eta_m \frac{d\epsilon}{dt} \tag{5.41}$$

and

$$P_2 = E_a \epsilon. \tag{5.42}$$

Since $P = P_1 + P_2$ we obtain from (5.41) and (5.42)

$$E_m P + \eta_m \frac{dP}{dt} = \eta_m (E_a + E_m)\frac{d\epsilon}{dt} + E_m E_a \tag{5.43}$$

Similarly, if in model 27 (b) we denote the stiffness of the springs by E_v and E_a' and the 'viscosity' of the dashpot by η_v and sum the strains in the auxiliary spring and the Voigt element in series with it, we again obtain a relation between the applied force P and the strain ϵ. This is:

$$(E'_a+E_v)P+\eta_v\frac{dP}{dt} = \eta_v E'_a\frac{d\epsilon}{dt}+E_v E'_a\epsilon. \qquad (5.44)$$

Now equations (5.43) and (5.44) are equivalent if $E'_a = E_a+E_m$,

$$E_v = E_a(1+E_a/E_m) \quad \text{and} \quad \eta_v E_m E_a = \eta_m E_v E'_a.$$

The mechanical behaviour of the two types of model is thus identical and, from a mathematical point of view, it is purely a matter of convenience which is used.

Propagation of stress waves in a visco-elastic solid

The theory of the vibrations of a Boltzmann solid which obeys equation (5.38) leads to extremely complicated mathematical analysis in that the solution of partial integro-differential equations is involved. V. Volterra (1931) in his theory of functionals has treated this problem, but the results of the theory have so far found very little application in the study of the dynamic behaviour of visco-elastic materials.

The much simpler problem of the vibration of specimens of Voigt and Maxwell solids has been discussed earlier in this chapter and we shall now consider the propagation of stress waves through such media. In deriving the relations between stress and acceleration at the beginning of Chapter II (equations (2.7)) no assumptions were made with regard to the stress-strain behaviour of the solid medium and these equations will therefore apply to the motion of a visco-elastic solid. As mentioned earlier in this chapter the relations between stress and strain for a Voigt solid are of the same form as those for an elastic solid if the operator $\lambda+\lambda'(\partial/\partial t)$ is used instead of λ and the operator $\mu+\mu'(\partial/\partial t)$ is used instead of μ. Thus on substituting for the stress components in equations (2.7) we now obtain for the equation of motion in the x-direction

$$\rho\frac{\partial^2u}{\partial t^2} = \left[(\lambda+\mu)+(\lambda'+\mu')\frac{\partial}{\partial t}\right]\frac{\partial\Delta}{\partial x}+\left(\mu+\mu'\frac{\partial}{\partial t}\right)\nabla^2u, \quad (5.45)$$

with similar equations for v and w. As with an elastic solid these relations lead to differential equations for irrotational and equivoluminal motion. Thus the equation for the propagation of the displacement u in a dilatational wave now becomes

$$\rho \frac{\partial^2 u}{\partial t^2} = (\lambda+2\mu)\,\nabla^2 u + (\lambda'+2\mu')\,\nabla^2\!\left(\frac{\partial u}{\partial t}\right) \qquad (5.46)$$

and for a distortion wave we have

$$\rho \frac{\partial^2 u}{\partial t^2} = \mu \nabla^2 u + \mu' \nabla^2\!\left(\frac{\partial u}{\partial t}\right). \qquad (5.47)$$

Similar equations are obtained for v and w.

Now if for simplicity we consider a plane distortional wave propagated in the x-direction with its particle motion in the z-direction, we have from (5.47)

$$\rho \frac{\partial^2 w}{\partial t^2} = \mu \frac{\partial^2 w}{\partial x^2} + \mu' \frac{\partial^3 w}{\partial x^2 \partial t}. \qquad (5.48)$$

This equation will not in general be satisfied by a solution of the type $w = F(x-ct)$ or $w = F(x+ct)$. If we try a harmonic solution so that w is the real part of $A\exp[i(pt-f_1 x)]$ and substitute in (5.48) we obtain

$$\rho p^2 = \mu f_1^2 + i\mu' f_1^2 p. \qquad (5.49)$$

In order that this equation should be satisfied f_1 must be complex and if we put it equal to $f+i\alpha$ and equate real and imaginary parts of (5.49) we find

$$\rho p^2 = \mu(f^2-\alpha^2)-2\mu' pf\alpha \qquad (5.50)$$

and

$$2\mu\alpha f = -\mu' p(f^2-\alpha^2),$$

so that

$$f^2 = \frac{\mu\rho p^2}{2(\mu^2+\mu'^2 p^2)}\left\{\left[1+\frac{(\mu' p)^2}{\mu^2}\right]^{\frac{1}{2}}+1\right\} \qquad (5.51)$$

and

$$\alpha^2 = \frac{\mu\rho p^2}{2(\mu^2+\mu'^2 p^2)}\left\{\left[1+\frac{(\mu' p)^2}{\mu^2}\right]^{\frac{1}{2}}-1\right\}.$$

This complex value of f_1 means that the wave is damped exponentially as it proceeds through the solid, and the expression for w may be written

$$w = \exp(\alpha x)\cos(pt-fx). \qquad (5.52)$$

(From equation (5.51) it may be seen that α may have either a positive or negative value. The positive root, however, does not correspond to the physics of the problem.) It may be seen that both f and α vary with the frequency, and when $\mu'p$ is small compared with μ, α becomes proportional to the square of the frequency. Now α is here the same as in equation (5.22) which gives the relation between the damping of progressive waves and the specific loss $\Delta W/W$. Equation (5.22) shows that α is proportional to the product of the frequency and the specific loss so that for low values of μ' the specific loss and hence the logarithmic decrement will be proportional to the frequency. This is in agreement with equation (5.6) obtained for a vibrating Voigt solid.

The relation for f in equation (5.50) shows that for small values of $\mu'p$, $f = p(\rho/\mu)^{\frac{1}{2}}$. Hence the phase velocity which is p/f becomes $(\mu/\rho)^{\frac{1}{2}}$ and is the same as for an elastic solid. Thus the phase velocity is less sensitive to frequency than the attenuation and only begins to change when p becomes comparable with μ/μ'. The treatment for dilatational waves is identical with that for distortional waves except that $(\lambda+2\mu)$ takes the place of μ and $(\lambda'+2\mu')$ takes the place of μ'.

The propagation of longitudinal waves along a thin rod of material which behaves like a Voigt solid gives the equation of motion

$$\rho \frac{\partial^2 u}{\partial t^2} = E \frac{\partial^2 u}{\partial x^2} + E' \frac{\partial^3 u}{\partial x^2 \partial t}, \qquad (5.53)$$

where E' is the 'extensional viscosity' as given by equations (5.29) and (5.30). This again leads to equations similar to (5.51) with E and E' taking the places of μ and μ' respectively.

Hillier (1949) has considered the propagation of longitudinal sinusoidal waves along a visco-elastic filament and has derived the relations for a Maxwell solid, a Voigt solid, and a solid which behaves like the models in Fig. 27. For a Maxwell solid the relation between stress and strain is

$$\frac{\partial \sigma}{\partial t} = E^* \frac{\partial^2 u}{\partial x \partial t} - \frac{\sigma}{\tau} \qquad (5.54)$$

(see equation (5.23)). τ is here the relaxation time and if we call η^* the 'equivalent viscosity' of the solid, $\tau = \eta^*/E^*$.

The equation of motion of the rod from Newton's second law is

$$\rho \frac{\partial^2 u}{\partial t^2} = \frac{\partial \sigma}{\partial x}, \tag{5.55}$$

so that differentiating (5.54) with respect to x and substituting for σ from (5.55) we obtain

$$\rho \frac{\partial^3 u}{\partial t^3} - E^* \frac{\partial^3 u}{\partial x^2 \partial t} + \frac{\rho}{\tau} \frac{\partial^2 u}{\partial t^2} = 0. \tag{5.56}$$

Equation (5.56) will not be satisfied by a solution of the type $u = A \exp i(pt - f_1 x)$ unless f_1 is complex and if, as before, we write $f_1 = f + i\alpha$ we find

$$
\begin{aligned}
f^2 &= [\rho p^2 / 2 E^*][(1 + p^{-2}\tau^{-2})^{\frac{1}{2}} + 1] \\
\text{and} \qquad \alpha^2 &= [\rho p^2 / 2 E^*][(1 + p^{-2}\tau^{-2})^{\frac{1}{2}} - 1]
\end{aligned}
\tag{5.57}
$$

When p is large compared with $1/\tau$, that is, when the period of the stress waves is short compared with the relaxation time, $f^2 = \rho p^2 / E^*$ so that the wave velocity is $(E^*/\rho)^{\frac{1}{2}}$, which is the same as that for an elastic rod of Young's modulus E^*, the attenuation factor α becomes $(\rho / 4 E^* \tau^2)^{\frac{1}{2}}$ and is therefore independent of frequency. The specific loss is proportional to α/p (see equation (5.22)) and is therefore inversely proportional to the frequency. This is in agreement with equation (5.37) for a vibrating Maxwell solid. The third type of model considered by Hillier is that shown in Fig. 27 (b), where an auxiliary spring is placed in series with the Voigt model. The stress-strain relation for such a model is given by equation (5.44) as

$$(E_a' + E_v)\sigma + \eta_v \frac{\partial \sigma}{\partial t} - \eta_v E_a' \frac{\partial \epsilon}{\partial t} - E_v E_a' \epsilon = 0,$$

where σ is the stress and ϵ is the strain.

If we differentiate this with respect to x and substitute for $\partial\sigma/\partial x$ from (5.55) we obtain

$$\rho \eta_v \frac{\partial^3 u}{\partial t^3} + \rho(E_a' + E_v) \frac{\partial^2 u}{\partial t^2} - \eta_v E_a' \frac{\partial^3 u}{\partial x^2 \partial t} - E_v E_a' \frac{\partial^2 u}{\partial x^2} = 0. \tag{5.58}$$

If we put η_v/E_v equal to τ, the 'time of retardation' of the Voigt element, and as before try a solution of the form

$$u = A \exp[i\{pt-(f+i\alpha)x\}],$$

we find that (5.58) will be satisfied if

$$
\begin{aligned}
f^2 &= \frac{\rho p^2}{2E_c\,E_a'}\left[\left(\frac{E_a'^2+E_c^2p^2\tau^2}{1+p^2\tau^2}\right)^{\frac{1}{2}}+\frac{E_a'+E_c\,p^2\tau^2}{1+p^2\tau^2}\right] \\
\text{and} \qquad \alpha^2 &= \frac{\rho p^2}{2E_c\,E_a'}\left[\left(\frac{E_a'^2+E_c^2p^2\tau^2}{1+p^2\tau^2}\right)^{\frac{1}{2}}-\frac{E_a'+E_c\,p^2\tau^2}{1+p^2\tau^2}\right]
\end{aligned} \right\}, \quad (5.59)
$$

where E_c is written for the modulus for the two springs in series, so that $1/E_c = 1/E_a'+1/E_v$. If E_a' is very large, i.e. if the auxiliary spring is very stiff, equations (5.59) simplify to equations of the form (5.51) for a simple Voigt solid. If on the other hand E_v is very small, equation (5.58) becomes identical with (5.56) for a Maxwell solid.

The velocity c of propagation of waves along the rod is given by p/f and from equations (5.59) it may be seen that for small values of p, c approaches $(E_c/\rho)^{\frac{1}{2}}$ whilst for very large values of p, c approaches $(E_a'/\rho)^{\frac{1}{2}}$. Thus at frequencies low compared with $1/\tau$, the reciprocal of the retardation time, the velocity of propagation corresponds to the elastic behaviour of the two springs in series, whilst at high frequencies the Voigt spring is inoperative and the velocity of propagation depends on the modulus of the auxiliary spring. The damping of the wave is given by α and increases with increasing frequency; the specific loss, however, is proportional to α/p and it may be seen from (5.59) that it approaches zero both for very small and very large values of $p\tau$, with a maximum between.

Fig. 28 shows the curves obtained on plotting the velocity of propagation and the specific loss against $p\tau$ for the special case $E_a' = E_v$. The results are plotted in non-dimensional form, the velocity being given as c/c_0, where c_0 is the velocity of propagation at 'zero' frequency so that $c_0^2 = E_c/\rho$, and the damping is given as $\alpha c_0/p$. This will be proportional to the specific loss in the solid. It may be seen from the figure that the damping is a maximum when $p\tau$ equals $1\cdot18$, and that at frequencies higher or lower than this the damping falls off rapidly. We may compare the velocity curve in Fig. 28 with the dispersion curves shown in Fig. 14 for longitudinal waves in an elastic cylindrical rod. The dispersion in the latter case is due to purely geometrical considerations, whilst we are here dealing with dispersion produced by the visco-elastic properties of the solid. It is of interest, however,

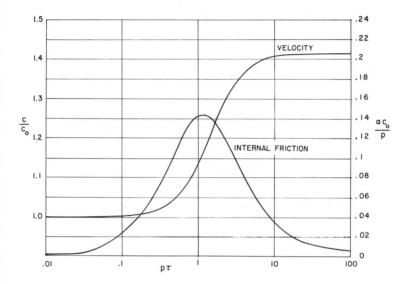

FIG. 28. The variation of velocity of propagation and damping loss with frequency in a solid which behaves like the model in Fig. 27 (b), with E_a' equal to E_v.

to note that the directions of the dispersion are opposite in the two cases, high-frequency waves travelling faster than low-frequency ones in a visco-elastic solid whilst the reverse is true for elastic waves in a cylinder the diameter of which is comparable with the wavelength. It should be of interest to investigate propagation in a visco-elastic cylinder the diameter of which is comparable with the wavelength as two opposing effects will then be present.

It should be emphasized that very few solids behave, even approximately, like either the Maxwell or the Voigt model and that it is only in terms of a relaxation time spectrum that the dynamic behaviour can be defined adequately. The only reason for using simpler models with single relaxation times is that the mathematics otherwise becomes extremely involved. When, however, the dynamic mechanical behaviour of a visco-elastic solid is required for only a restricted region of frequencies the 'elasticity' and 'viscosity' obtained by using a single Maxwell or Voigt element is often a convenient method of describing its mechanical properties under the prescribed conditions.

The memory function and the associated relaxation time spectrum describe the mechanical behaviour of a solid so long as it remains a linear system, that is so long as there is a linear relation between stress and strain when time is fixed. This is implied in the superposition principle, and for most solids there appears to be a value of strain up to which this is at least approximately true. For strains which exceed this value the treatment will no longer apply. The theory of stress propagation of non-linear systems has not yet been developed except in one or two rather special cases. These will be considered in Chapter VII, where the propagation of plastic waves and shock waves is discussed.

Mechanism of internal friction

Internal friction in solids may be produced by several different mechanisms, and although ultimately these all result in the mechanical energy being transformed into heat, two different dissipative processes are involved. These two processes are roughly the counterparts of viscosity losses and thermal conduction losses in the transmission of sound waves through fluids.

The first type of process depends directly on the anelastic behaviour of the solid, and if the stress-strain curve for a single cycle of the vibrations is in the form of an hysteresis loop the area enclosed by this loop will represent mechanical energy which has been lost in the form of heat. In taking a specimen round a stress-cycle 'statically' a certain amount of energy is dissipated and this will appear as part of the specific loss when the specimen is vibrating. As pointed out by Gemant and

Jackson (1937), even when this hysteresis loop is too small to be measured statically it may have a considerable effect on the damping of vibrations since a specimen may be taken round a very large number of hysteresis cycles in a single vibration experiment. The effect of this type of hysteresis is that the energy loss per cycle will be constant so that the specific loss and hence the logarithmic decrement will be independent of frequency. Gemant and Jackson have found that for many materials the logarithmic decrement is in fact constant over a considerable range of frequency, and conclude that the principal cause of internal friction in these cases may be associated simply with the 'static' non-linear stress-strain behaviour of the materials. Similar results have been obtained by Wegel and Walther (1935) at higher frequencies.

In addition to static hysteresis, many materials show losses which are associated with the velocity gradients set up by the vibrations, and the forces producing these losses may be considered to be of a viscous nature. As we have seen, such forces imply that the mechanical behaviour will depend upon the rate of straining and this effect is particularly marked with organic long chain polymers; the subject of Rheology is mainly concerned with such time dependence.

It is possible to distinguish two types of viscous loss in solids and these correspond qualitatively to the behaviour of the Maxwell and Voigt models described in the previous section. Thus when a load is maintained it may result in irrecoverable deformation as in the Maxwell model, or the strain may approach a constant value asymptotically with time and recover slowly when the load is removed, as it does in the Voigt model. This latter type of viscosity has sometimes been called internal viscosity, and the mechanical behaviour of such solids is sometimes referred to as *retarded elasticity*.

On a molecular scale the explanation of viscous effects in solids is not at all well understood, mainly because the types of microscopic processes which result in the dissipation of mechanical energy into heat are still largely a question of conjecture. Tobolsky, Powell, and Eyring (1943) and Alfrey (1948) give accounts of a treatment of visco-elastic behaviour in terms of *rate process theory*. The assumption in this treatment is that each molecule (or in the case of long chain polymers, each segment of a molecular chain) performs thermal oscillations in an 'energy well' produced by its neighbours. As a result of thermal fluctuations it will occasionally have sufficient energy to leave the well and in the absence of external forces diffusion will take place equally in all directions. The diffusion rate is dependent on the probability of the molecule receiving sufficient energy to leave the well, and hence on the absolute temperature of the solid. If a hydrostatic pressure is applied to the solid, the height of the energy well will change and the diffusion rate will alter but will still remain equal in all directions. If a uniaxial tension is applied, however, the height of the well will be lower in the direction of the tensile stress than it is at right angles to this direction; consequently molecules will be more likely to travel in the direction parallel to the tensile stress than perpendicular to it, and there will be a net movement away from the well in these directions. The effect of such flow will be to convert the elastic energy stored in the solid into random thermal motion which on the macroscopic scale will appear as internal friction. Where whole molecules are moving the flow will be irrecoverable and the behaviour will be analogous to the Maxwell model, whilst where segments of molecules are involved the material will behave like the Voigt model and exhibit retarded elasticity.

If certain assumptions are made about the shape of the potential energy well and the nature of the molecular groups which are oscillating in it, it may be shown (Tobolsky, Powell, and Eyring, 1943, p. 125) that the theory leads to mechanical behaviour of the solid which is similar to that described by the spring and dashpot models discussed earlier in this chapter. The dependence of the visco-elastic properties on temperature is emphasized in this type of treatment and thermodynamic relations can be developed from it. The main disadvantages in applying the theory to real solids quantitatively lies in the fact that the nature of the potential well for solids is largely a matter of surmise and that often many different processes may be taking place simultaneously. Nevertheless it is so far almost the only serious approach to a molecular explanation of the observed effects and it provides a useful basis for future developments.

For homogeneous non-metallic solids the losses appear to be chiefly of the type described above, and the internal friction is associated with the anelastic behaviour of the material rather than with its macroscopic thermal properties. With metals, however, it is the thermal losses which are generally more important, and Zener (1948) discusses several different thermal mechanisms which will result in the dissipation of the mechanical energy into heat.

Changes in the volume of a solid will be accompanied by changes in temperature; thus, when a solid is compressed its temperature will rise whilst when it is extended its temperature will fall. For simplicity we may consider the case of a reed vibrating in flexure. Each time the reed is bent the inner side will be heated whilst the outer side will be cooled, so that there will be a continual flow of heat back and forth across the reed as it performs flexural vibrations. If the motion takes place very slowly the heat transfer will take place *isothermally* and hence reversibly, so that there will be no loss at very low frequencies of vibration. When the vibrations take place so rapidly that the heat has no time to flow across the reed the conditions will be adiabatic and again there will be no loss. For flexural vibrations whose periods are comparable with the time it takes for heat to flow across the reed there will be an irreversible conversion of mechanical energy into heat which will appear as internal friction. Zener (1937) shows that for a vibrating reed the specific loss is given by

$$\frac{\Delta W}{W} = 2\pi \frac{(E_S - E_T)}{E_S} \frac{N_0 N}{N_0^2 + N^2}, \qquad (5.60)$$

where E_S and E_T are the adiabatic and isothermal values of Young's modulus for the material, N is the frequency of vibration, and N_0 is the relaxation frequency which, for a reed of rectangular cross-section, is given by

$$N_0 = \pi K'/(2C_p \rho d^2), \qquad (5.61)$$

K' being the thermal conductivity, C_p the specific heat at constant pressure, ρ the density, and d the thickness of the reed in the plane of vibration.

Bennewitz and Rötger (1938) have measured the internal friction of German silver reeds in transverse vibration. Their experimental results are shown in Fig. 29 together with the

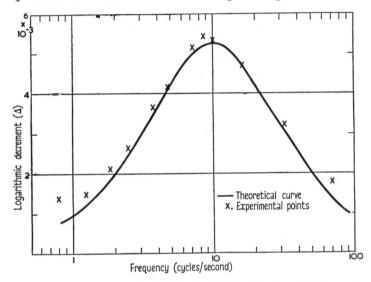

FIG. 29. Comparison between the values of internal friction of German silver in transverse vibration as measured by Bennewitz and Rötger and as given by Zener's theoretical relation.

theoretical curve obtained from equation (5.60). No arbitrary parameters were used in deriving this curve and the agreement between theory and experiment is remarkably good. In the region of frequencies around N_0 (c. 10 c./sec.) thermal conduction across the reed is thus clearly the principal cause of internal friction. At frequencies remote from N_0 it may be seen that the experimental values of internal friction are higher than those predicted by the theory and this indicates that other effects are here becoming relatively more important. Longitudinal stress will produce similar effects since parts of the specimen are compressed whilst others are extended, and in this case thermal currents will flow parallel to the direction of propagation. Since the distance between compressions and rarefactions in this case will be one-half of a wavelength, the losses due to this cause will be small at ordinary frequencies.

The above type of thermal loss will apply whether or not the solid is homogeneous. When the material is not homogeneous there are additional mechanisms which result in thermal losses. Thus with a polycrystalline material neighbouring grains may have different crystallographic directions with respect to the direction of the strain and will therefore be stressed by different amounts when the specimen is deformed. The temperature will thus differ from crystallite to crystallite and minute thermal currents will flow across the crystal boundaries. As in the case of the thermal conduction losses in a vibrating cantilever there will be a low-frequency limit when the deformations take place so slowly that the volume changes are isothermal and no energy is lost, and an upper frequency limit when the deformations take place so rapidly that the heat does not have time to flow away during an oscillation and there is again no loss. The maximum loss occurs when the applied frequency falls between these two limits, and the value of this frequency will depend on the crystal grain size and on the thermal conductivity of the medium. Zener has derived a relation for the frequency at which this loss is a maximum. This equation is similar to (5.61) and is

$$N_0 = 3\pi K'/(\rho C_p a^2), \qquad (5.62)$$

where a is the average linear grain size.

Randall, Rose, and Zener (1939) have measured the internal friction in brass specimens of various grain sizes and found that, at the frequencies used, maximum damping occurred when the size of grain had a value very close to that predicted by equation (5.62). The magnitude of the internal friction due to these microscopic thermal currents will depend on the type of crystal structure as well as on the crystal size, and will increase with increasing elastic anisotropy of the individual crystallites. Zener (1948, p. 80) suggests that at very high frequencies this thermal flow is confined almost entirely to the immediate vicinity of the grain boundaries; this leads to the relation that the specific loss is proportional to the square root of the frequency of vibration. This result has been confirmed experimentally for brass by Randall, Rose, and Zener (1939). At very low frequencies, on the other hand, the heat flow takes place throughout the material and this leads to the relation that internal friction is proportional to the first power of the frequency. The experimental results of Zener and Randall (1940) are in agreement with this.

There are two other types of thermal loss which should be mentioned. The first is thermal conduction to the surrounding air; the rate of loss from this cause is so slow, however, that it will only have an effect at very low frequencies of vibration. The other type of loss might arise from the absence of thermal equilibrium between the Debye normal modes, and this loss would be analogous to the damping of ultrasonics in gases resulting from the finite time necessary for the thermal energy to become divided between the various degrees of freedom of the gas molecules. In solids, however, equilibrium between the separate modes is established so rapidly that internal friction due to this cause would be expected to occur only at frequencies of the order of 1,000 megacycles per second. The theory of this effect has been discussed by Landau and Rumer (1937) and more recently by Gurevich (1945).

For polycrystalline metals Kê (1947) has investigated internal friction produced by 'viscous slip' at the crystal boundaries. He carried out experiments on the damping of torsional oscillations in pure aluminium, and has shown that the internal friction in this case could be adequately accounted for by assuming that the metal at the boundaries of the crystals behaves in a viscous manner.

There are two other processes which occur in crystalline solids when they are deformed which would be expected to result in internal friction. The first of these is the movement in the crystals of regions of disarray which are known as *dislocations*. The other process is the ordering of solute atoms on the application of stress; this occurs when there is an impurity dissolved in the crystal lattice. The role of dislocations in the plastic deformation of crystals was first discussed by Orowan (1934), Polanyi

(1934), and Taylor (1934), and although it seems probable that the movements of such dislocations are often an important cause of internal friction, especially for large strains, the exact mechanism by which the elastic energy is dissipated is not at present clear (see Bradfield, 1951). The influence of dissolved impurities in a crystal lattice on internal friction was first discussed by Gorsky (1936) and later by Snoek (1941). The reason that the presence of such solute atoms results in internal friction is that the equilibrium distribution of solute atoms in the stressed crystal is different from the equilibrium distribution when the crystal is unstressed. When a stress is applied the new equilibrium takes time to be established so that the strain lags behind the stress. This introduces a relaxation process which is of importance for oscillating stresses the periods of which are comparable with the relaxation time involved. The rate at which equilibrium is set up will depend very markedly on the temperature, so that this type of internal friction should be highly temperature sensitive.

A particular case of internal friction is that found in ferromagnetic materials. Becker and Döring (1939) give a comprehensive review of the experimental and theoretical work for materials of this type, the problem being of importance in the application of the magnetostrictive effect in the generation of ultrasonics. It is found that the internal friction in ferromagnetic materials is very much greater than for other metals and decreases when they are magnetized; it also decreases rapidly with increasing temperature as the Curie point is approached.

Another mechanism which attenuates stress waves in solids but which is not strictly internal friction is scattering. This occurs in a polycrystalline material when the wavelength becomes comparable with the grain size, and Mason and McSkimin (1947) have made measurements of this effect in aluminium rods and have shown that when the wavelength is long compared with the grain size the attenuation (α) is inversely proportional to the fourth power of the wavelength. This relation is the same as that given by Rayleigh (1894, vol. ii, p. 194) for the scattering of sound in gases.

CHAPTER VI

EXPERIMENTAL INVESTIGATION OF DYNAMIC ELASTIC PROPERTIES

In the last chapter various methods of defining internal friction were considered, and in order to do this it was necessary to outline the types of experiment by which this quantity can be measured. In the present chapter some of the methods which have been used to investigate the dynamic elastic behaviour of solids will be described in greater detail and the experimental results obtained with them will be reviewed.

The methods which have been employed may be divided into several distinct classes, and these are:

(i)　free vibrations;

(ii)　resonance methods;

(iii)　wave propagation methods;

(iv)　direct observation of stress-strain curves.

The type of method used depends on the period and amplitude of the deformation which it is required to investigate and on the shape of specimen which is readily available, and some workers have used several or all of the above methods in a single investigation. For convenience however the methods will be discussed here separately.

Free vibrations

If the mechanical properties of the material under investigation are linear, i.e. if its elastic properties are independent of amplitude, then at a given frequency of oscillation the period and logarithmic decrement of free oscillations will define its mechanical behaviour. The experimental technique for this type of measurement is simple and a large number of investigations of internal friction have been carried out by this method. Since in order to make observation easy a large amplitude is desirable, the method has been employed mainly with torsional and flexural oscillations. With very slow oscillations both the period and logarithmic decrement can be measured directly, whilst at higher frequencies a photographic or electrical method may be used for recording. In order to cover a range of frequency, specimens of different sizes may be employed. It is generally more convenient, however, to use auxiliary inertia members so that the period of oscillation can be varied with the same specimen.

The earliest measurements by this method were made by Weber (1837), who used the specimen as the suspension element in a ballistic galvanometer. In recent years the free oscillation method has been employed by Föppl (1936), Gemant and Jackson (1937), Norton (1939), Kê (1947), Guillet (1948), Kamel (1949), Wert (1949), and Lethersich (1950). Föppl, Norton, Guillet, and Kê have used the torsional method with metallic specimens, Kamel measured the damping in transverse vibration of specimens of metals and glass, whilst Wert used longitudinal oscillations with single crystal zinc specimens. Gemant and Jackson used both the torsional and the flexural method for metals, glass, and dielectrics and Lethersich has investigated several plastics by the torsional method.

As was shown in the last chapter, the losses in metals are largely thermal, and the Zener theory explains most of the observed results. The internal friction of plastics and dielectric materials is generally very much higher, and the value of the elastic modulus often changes very rapidly with frequency. These effects will be discussed later in the chapter when other methods of measuring dynamic elastic properties have been described, but it may be useful here to have an estimate of the relative values of internal friction in various materials, and this is shown in Table I which gives the values obtained by Gemant and Jackson at frequencies between 0·3 and 10 cycles per second.

The lowest recorded values of internal friction are those of piezo-electric quartz resonators. Van Dyke (1938) reports a value of $1·5 \times 10^{-6}$ for the logarithmic decrement of a quartz ring resonating at 100 kilocycles per second; it may be seen that the highest value in Table I is more than 50,000 times as great as this.

TABLE I

Logarithmic decrement for various materials as measured by Gemant and Jackson

Material	Torsion	Flexure
Steel . .	. .	$1 \cdot 7 \times 10^{-3}$
Copper .	. .	$3 \cdot 2 \times 10^{-3}$
Fused quartz	$2 \cdot 6 \times 10^{-3}$	. .
Glass, lead .	$4 \cdot 2 \times 10^{-3}$	. .
Glass, soft .	$14 \cdot 0 \times 10^{-3}$	$9 \cdot 5 \times 10^{-3}$
Wood . .	. .	$27 \cdot 0 \times 10^{-3}$
Ebonite .	$29 \cdot 0 \times 10^{-3}$	$85 \cdot 0 \times 10^{-3}$
Polystyrene .	$48 \cdot 0 \times 10^{-3}$	. .

A very elegant variation of the free oscillation method is due to Le Rolland (1931), who uses the specimen under investigation to couple two pendulums of identical period. When one pendulum is set in motion its oscillations are found to decrease slowly whilst the other oscillates with increasing amplitude. With a perfectly elastic specimen the first pendulum eventually stops oscillating and the process is reversed, so that each pendulum in turn acquires all the kinetic energy. The time between two stops for either pendulum may be shown to be approximately proportional to the force required to displace the end of the specimen to which the pendulums are attached through unit distance, and this gives a measure of the elastic modulus for the material under the effect of the oscillating force. By using compound pendulums the period of the oscillations may be varied between a fraction of a second and several seconds, and the variation of the dynamic modulus with frequency may hence be determined.

As pointed out by Le Rolland (1948) this apparatus may also be conveniently employed to investigate the internal friction in the specimen. When the two pendulums are oscillating in opposing phase there is no net lateral force on the specimen and the oscillations will decay purely as a result of air resistance and friction at the supports. When the two pendulums oscillate in phase the air damping and friction at the supports will still be present, but in addition the specimen is taken round a stress cycle during each oscillation. The energy loss within the specimen will result in higher damping, and the difference between the rate of decay of oscillations for the two modes gives a measure of the internal friction of the specimen.

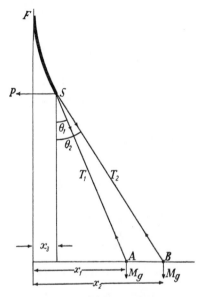

FIG. 30. Le Rolland pendulums.

The type of elastic deformation which the specimen undergoes will depend on the way in which the pendulums are attached to it. The simplest type of arrangement is to have the specimen in the form of a rod, one end of which is clamped rigidly whilst the other holds the support from which the pendulums swing in parallel planes. This is shown diagrammatically in Fig. 30. With such an arrangement the specimen is bent by the action of the pendulums so that Young's modulus is the elastic constant which is relevant. With an alternative method of support torsional oscillations of the specimen may be induced and the rigidity modulus may be measured.

In considering the equations of motion of the arrangement shown in Fig. 30 the mass of the specimen FS and of the support S of the two pendulums SA, SB is taken to be negligible compared with that of the bobs of the pendulums, which are each of mass M and are suspended by threads of length l. At any time t let the horizontal displacement of the support S be x_3 and the horizontal displacements of A and B be x_1 and x_2 respectively. Let the tension in the threads supporting the bobs be T_1 and T_2 respectively. The equation of motion of the pendulum SA then becomes

$$M\frac{d^2x_1}{dt^2} + T_1 \sin\theta_1 = 0, \tag{6.1}$$

where θ_1 is the angle AS makes with the vertical through S. If this angle is small we have $\sin\theta_1 = (x_1-x_3)/l$, and $T_1 = Mg$, so that (6.1) becomes

$$M\frac{d^2x_1}{dt^2} + Mg\frac{(x_1-x_3)}{l} = 0,$$

or $$\frac{d^2x_1}{dt^2} + \frac{g}{l}(x_1-x_3) = 0. \tag{6.2}$$

Similarly for the second pendulum

$$\frac{d^2x_2}{dt^2} + \frac{g}{l}(x_2-x_3) = 0. \tag{6.3}$$

Now resolving the forces acting on S horizontally we have

$$T_1 \sin\theta_1 + T_2 \sin\theta_2 = P, \tag{6.4}$$

where P is the horizontal restoring force set up by the specimen; for an elastic specimen this will be proportional to x_3, the displacement from equilibrium, so that P may be written as $A'x_3$, where A' is a quantity depending on Young's modulus and the dimensions of the specimen. Thus if θ_1 and θ_2 are small (6.4) becomes

$$Mg\frac{x_1-x_3}{l} + Mg\frac{x_2-x_3}{l} = A'x_3,$$

or $$x_3 = \frac{Mg}{A'l+2Mg}(x_1+x_2) = B'(x_1+x_2), \tag{6.5}$$

where $B' = Mg/(A'l+2Mg)$.

Now if we subtract corresponding sides of equation (6.2) from (6.3) we obtain

$$\frac{d^2}{dt^2}(x_1-x_2) + \frac{g}{l}(x_1-x_2) = 0,$$

and the general solution of this is

$$(x_1 - x_2) = C_1 \sin(p_1 t + \alpha_1), \tag{6.6}$$

where $p_1^2 = g/l$, and C_1 and α_1 depend on the initial conditions. Thus $(x_1 - x_2)$, the distance between the two pendulums, varies sinusoidally with time, the period being the natural period of a single pendulum.

If we add (6.2) and (6.3) and substitute from (6.5) we obtain

$$\frac{d^2}{dt^2}(x_1 + x_2) + \frac{g}{l}(1 - 2B')(x_1 + x_2) = 0,$$

and the general solution of this is

$$(x_1 + x_2) = C_2 \sin(p_2 t + \alpha_2), \tag{6.7}$$

where $p_2^2 = (g/l)(1 - 2B')$ and C_2 and α_2 are constants depending on the initial conditions. From (6.6) and (6.7) we finally obtain

$$\left. \begin{aligned} 2x_1 &= C_1 \sin(p_1 t + \alpha_1) + C_2 \sin(p_2 t + \alpha_2) \\ 2x_2 &= C_2 \sin(p_2 t + \alpha_2) - C_1 \sin(p_1 t + \alpha_1) \end{aligned} \right\}. \tag{6.8}$$

Now if pendulum A is displaced and released at zero time we have that at $t = 0$, $x_2 = \dot{x}_2 = \dot{x}_1 = 0$, and this leads to $C_1 = C_2$ and $\alpha_1 = \alpha_2 = \pi/2$, so that

$$x_1 = C_1 \cos\frac{p_1 + p_2}{2} t \cos\frac{p_1 - p_2}{2} t$$

and

$$x_2 = C_2 \sin\frac{p_1 + p_2}{2} t \sin\frac{p_1 - p_2}{2} t.$$

These expressions correspond to oscillations of frequency $(p_1 + p_2)/4\pi$ the amplitudes of which are modulated with a frequency $(p_1 - p_2)/4\pi$, the latter being the *beat* frequency. Now if $(p_1 - p_2)$ is small compared with p_1, $(p_1 - p_2) \simeq (p_1^2 - p_2^2)/2p_1$ and from the expressions for p_1 and p_2 given for (6.6) and (6.7) we have

$$(p_1 - p_2) \simeq B'\sqrt{(g/l)}. \tag{6.9}$$

Now A' will generally be large compared with $2Mg/l$ so that from (6.5) the beat frequency will be approximately inversely proportional to A' which by definition is the force necessary to displace the end of the specimen through unit distance. It may be seen from equations (6.8) that the displacements of each of the pendulums is the sum of two separate sinusoidal oscillations of different frequencies, the relative amplitudes of which depend on the initial conditions. The two terms on the right-hand side of each equation correspond to the two different *modes* of the system, the first term being the *asymmetric mode* whilst the second is the *symmetric mode*.

The simple theory as given above has assumed that the specimen obeys Hooke's law and that the displacement of the end of the specimen is proportional to the force applied to it. If the specimen behaves in a visco-elastic manner there will be an additional term in equation (6.5) involving dx_3/dt and although (6.6) will still apply, (6.7) will now contain an exponential damping factor. Thus the symmetric mode will slowly decay whilst the asymmetric mode will be undamped except for air resistance. When one pendulum is set in motion under these conditions the oscillations will not stop completely, but will pass through a minimum. The ratio of the minimum to the maximum amplitudes gives a measure of the viscous loss in the specimen, and this fact has been used by Kovacs (1948) for measuring the internal friction of plastics.

As mentioned above, the main advantage of the free oscillation method is its simplicity, and the method is particularly suitable for work at low frequencies with specimens that have a low internal friction. By the use of photographic methods of recording, however, higher frequencies have been used and Lethersich (1950) has worked up to 1,000 cycles per second by this means. The main error in measurements of internal friction is in avoiding extraneous losses due to air damping, friction at the supports, etc. Where the internal friction is low this often results in large errors. As far as air damping is concerned this has sometimes been eliminated by working *in vacuo* as in Kamel's experiments, or alternatively a separate set of measurements can be made to allow for it as in the Le Rolland method.

Resonance method

The principle of this method of measuring elastic properties of materials is that if an oscillating force whose amplitude is fixed but whose frequency can be varied is applied to a mechanical system, the amplitude of the resulting vibration passes through a maximum at a frequency which is known as the resonant frequency of the system. The value of this resonant frequency depends on the elastic properties of the system, whilst the breadth of the resonance peak gives a measure of the dissipative forces which are present. It was shown in the last chapter that when the dissipative forces are large they change the value of the resonant frequency, but this effect can be allowed for when the value of the damping is known.

If suitable precautions are taken to eliminate extraneous damping due to air resistance, loss at the supports, etc., both the internal friction and the elastic constants of a specimen may be determined by this method and measurements have been made using longitudinal, flexural, and torsional oscillations at frequencies from a few cycles to several megacycles per second. The method may be used when the damping is so large that free oscillations decay too rapidly for accurate measurements to be made; on the other hand the method is not very suitable for specimens with very low internal friction since the resonance peak then becomes too sharp for accurate work. When a calibrated electronic oscillator is employed to drive the specimen it is unnecessary to make absolute measurements of the mechanical vibrations, and all that is required are relative values of the amplitude at frequencies around resonance. The main disadvantage of the resonance method is that the coupling between the driving system and the specimen may result in a change in resonant frequency and in the shape of the resonance peak. It is sometimes necessary to carry out a series of measurements with varying degrees of coupling to allow for this effect.

Quimby (1925) was among the first to employ the resonance method for measuring internal friction in solids. He used a piezoelectric quartz crystal to produce longitudinal oscillations in his specimens, which were in the form of rods. The crystal was cemented to one end of the specimen whilst the amplitude of the vibration was measured with a Rayleigh disk which was suspended near the other end. He worked with specimens of copper, aluminium, and glass at frequencies around 40 kilocycles per second. In later work Quimby (1932), Zacharias (1933), and

Cooke (1936) used this method to investigate losses in ferromagnetic materials.

Wegel and Walther (1935) employed both longitudinal and torsional oscillations with cylindrical rods of metals at frequencies between 100 and 10,000 cycles per second. The oscillations were generated electromagnetically by the eddy currents induced at one end of the specimen and the amplitude was measured by the current induced in a coil which vibrated in a stationary magnetic field at the other end of the specimen. Randall, Rose, and Zener (1939) used a similar method in their investigation on the relation between internal friction and grain size; the results of this work were discussed in the last chapter.

Bancroft and Jacobs (1938) used an electrostatic method of generating longitudinal oscillations in metal bars, the amplitude being detected with a condenser microphone, and a similar method has been used by Parfitt (1949) with high polymers at frequencies between 5 and 60 kilocycles per second. Bordoni (1947) has also described an electrostatic generator with a condenser microphone to detect the vibrations. The condenser microphone operates in a radiofrequency oscillator and Bordoni claims that by using a frequency modulation detector he can measure displacements in which the average movement of the surface is only a fraction of one Ångström unit.

One of the difficulties in the measurement of internal friction by the resonance method is loss of energy at the supports, and in most of the investigations mentioned the specimen was suspended by fine wires or threads. Even so, some energy will travel along the suspension and Gemant (1940) used the wire supports to excite and detect flexural oscillations. In his apparatus the specimen was in the form of a hollow metal cylinder suspended by two fine wires each of which was attached to the centre of the diaphragm of an earphone. One of the earphones which acted as generator was connected to an oscillator whilst the electrical output of the other was used for measuring the amplitude of the vibrations. Gemant measured the internal friction of paraffin wax by finding first the loss in the metal tube when empty and then when filled with the wax.

James and Davies (1934) have used the method of flexural vibrations for measuring the elastic constants of metal rods, and in another paper Davies and James (1934) have treated the effect of coupling between the generator and the specimen theoretically. In the experimental work the oscillations in the specimen were excited by means of an electromagnet which had two sets of windings. Through one of these windings a direct current was passed producing a static magnetic field whilst the alternating current passed through the other. The coupling depended on the value of the static magnetic field and by taking resonance curves with different values of the D.C. current the resonant frequency for zero coupling could be extrapolated. In a recent paper Hillier (1951) has used this method for measuring the dynamic value of Young's modulus in some high polymers. In this method the amplitude of oscillation is observed directly by means of a microscope with a calibrated eyepiece.

Nolle (1948) has investigated the elastic properties of rubber-like materials at frequencies between 0·1 cycles per second and 120 kilocycles per second and has employed five different experimental methods to cover this range of frequency. At the lowest frequencies, 0·1 to 25 cycles per second, he used a free oscillation method in which the rubber specimen acted as the elastic restoring force for a beam rocking on a knife edge. At the higher frequencies three different resonance methods and a wave-propagation method were employed. The wave-propagation method will be discussed in the next section, but we shall here mention briefly the resonance methods which Nolle describes. At frequencies between 10 and 500 cycles per second Nolle used a *vibrating reed* resonance method in which the specimen was clamped in a gramophone recording head and the flexural vibration was communicated through the clamp. This method is convenient, but suffers from the limitation that the frequencies which it will cover are restricted both by the mechanical response of the recording head and by the elastic properties of the specimen, since the value of the resonant frequency can be changed only by altering the size or shape of the specimen.

At frequencies of the order of a few hundred cycles per second

Nolle used a longitudinal resonance method in which a strip of the rubber was clamped between two gramophone cutting heads, one of which acted as generator whilst the other was the receiver. Here again the resonant frequencies depended on the length of the specimen, although a series of harmonics could be investigated with a single sample. The limitation of this method is that it assumes that the supports of the specimen are much more rigid than the specimen itself, and although this assumption is generally valid for rubber it is not justified with specimens of harder materials.

At the highest frequencies (12–120 kc./sec.) Nolle used a magnetostriction resonance method in which the sample of rubber was held against a nickel rod which was driven magnetostrictively. The presence of the sample shifted the resonant frequency of the rod slightly and broadened the resonance peak. The shift in the resonant frequency gives a measure of the elasticity of the specimen whilst the increase in breadth of the resonance peak depends on the internal friction. Nolle does not claim a very high accuracy for this method, which may give variations in the results of 10 to 20 per cent., but the method has the advantage that very small specimens may be used.

Before leaving methods of measuring dynamic elastic properties which depend on forced oscillation the revolving-rod apparatus devised by Kimball (1941) should be mentioned. This differs in principle from the resonance methods described above and has been used to measure internal friction at frequencies between one cycle per second and several thousand cycles per second. The apparatus is shown diagrammatically in Fig. 31. The specimen is in the form of a cylindrical rod R, which is rotated by the pulley P. Near the end of the rod a bearing B is fitted from which a mass M is suspended; this deflects the rod vertically. When the rod rotates it is taken through a succession of flexural stress cycles, and the internal friction in the rod results in the strain lagging behind the stress. This causes the end of the rod to be deflected horizontally and the magnitude of the horizontal deflexion gives a measure of this phase lag and hence of the internal friction.

Kimball has used this method for measuring the logarithmic decrement of several metals, glass, wood, celluloid, and rubber. He found that with these materials, in the frequency range he employed, the logarithmic decrement was independent of frequency, or in other words, the energy loss per stress cycle was independent of the rate of loading. As mentioned earlier

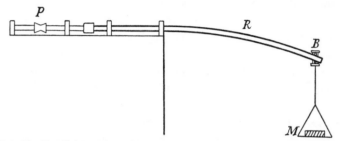

FIG. 31. Revolving-rod method for the measurement of internal friction (Kimball).

Gemant and Jackson (1937) obtained similar results by the free oscillation method and suggested that the mechanism of damping was a frictional rather than a viscous one. As shown in the last chapter, an alternative explanation of this effect is that in the frequency range which is being investigated the distribution of relaxation times is flat.

Wave-propagation methods

As was shown in the first part of this monograph there are a number of different types of elastic wave which can be propagated through solid media. In extended solids there are only two types of wave, and these are termed dilatational and distortional. Along solid rods three types of wave, extensional, torsional, and flexural can be propagated, whilst in plates extensional and flexural waves can be propagated. In addition to these, Rayleigh waves can travel along the surface of a solid so long as their wavelength is not large compared with the lateral dimensions of the specimen.

The velocities of propagation of all these elastic waves depend, amongst other factors, on the elastic constants and density of the solid, so that the dynamic elastic constants can be determined from the velocity of propagation. When the solid is not perfectly elastic some of the energy of the stress wave is dissipated as it passes through the medium, and as shown in Chapter V the magnitude of this attenuation can be correlated with the internal friction as determined in other ways. Some measurements of the velocity of propagation and attenuation of sinusoidal waves have been made at low frequencies with specimens in the form of strips or filaments and Young's modulus is here the operative elastic constant. At high frequencies dilatational and distortional pulses have been propagated through blocks of the material. The advantages that the wave propagation method has over the other methods which have been described are, first, that a range of frequencies can be covered with a single specimen, secondly that in measuring internal friction by this method it is easier to reduce extraneous losses at supports, and lastly that in non-dispersive media the method is capable of an extremely high degree of accuracy. Bradfield (1950) states that the elastic constants of metals can be measured to one part in 4,000 with ultrasonic pulses.

The disadvantages of wave-propagation methods are, first that the apparatus involved is generally more complicated than that used for resonance or free oscillation work, secondly that it is not always easy to ensure that a particular type of wave is being generated, and lastly that in dispersive media the interpretation of the results obtained with pulses is often difficult. This last point will be discussed more fully later in this section.

The work on the propagation of low-frequency longitudinal waves in filaments has been mainly concerned with the dynamic behaviour of rubber-like materials and high polymers, and the method has been used by Ballou and Silverman (1944), Nolle (1948), Ballou and Smith (1949), Witte, Mrowca, and Guth (1949), Hillier and Kolsky (1949), and Hillier (1950). Fig. 32 shows a block diagram of the type of apparatus used in this work. A variable frequency oscillator is connected to the driving unit. This can be a piezo-electric crystal, a magnetostrictive rod, or a loudspeaker, depending on the frequency and power required. The end of the filament is attached to the vibrating part of the driving unit in such a way that longitudinal waves are generated in the filament. The detector is a piezo-electric crystal which is connected to the filament through a small probe and can be moved along the specimen. The amplified electrical output from the detector and that from the driving unit are fed to a cathode-ray oscillograph, where the phases and amplitudes of the two sinusoidal vibrations are compared. From a series of measure-

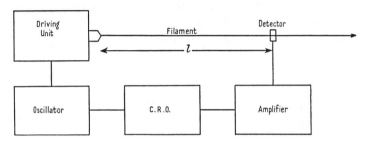

Fɪɢ. 32.

ments of phase difference against distance the wave velocity in the filament can be determined, whilst the attenuation can be obtained from the variation of amplitude with the distance between generator and detector.

If the detector does not reflect any of the incident stress wave and if the filament is sufficiently long for the wave reflected from its end to be of negligible amplitude the phase and amplitude relations are simple, since the longitudinal stress σ in the filament at distance x from the driving unit is given by

$$\sigma = \sigma_0 \exp(-\alpha x)\sin[p(t-x/c)],$$

where σ_0 is the stress amplitude at the driving unit end of the filament, p is 2π times the frequency, α is the attenuation factor which is related to the specific loss by equation (5.22) of Chapter V, and c is the wave velocity. Thus if the stress amplitude at distance x along the filament is $\hat{\sigma}$ and the phase difference between the driver and the detector is θ radians we have

$$\hat{\sigma} = \sigma_0 \exp(-\alpha x) \tag{6.10}$$

and

$$\theta = \frac{px}{c}. \tag{6.11}$$

Now although it is possible to eliminate the effect of reflection from the end of the filament, either by having it sufficiently long for the attenuation of the material to reduce the reflected wave to negligible amplitude or by terminating the end of the filament mechanically in such a way that the reflection is very small, the reflection at the detector is generally appreciable. Except when the attenuation in the material is very high this reflection at the detector affects both the phase and amplitude measurements, and if the amplitude of the reflected wave is m times that of the incident wave we have, instead of the relations given by (6.10) and (6.11),

$$\frac{\hat{\sigma}}{\sigma_0} = \frac{(1-m)\exp(-\alpha x)}{[1-2m\exp(-2\alpha x)\cos(2px/c)+m^2\exp(-4\alpha x)]^{\frac{1}{2}}} \tag{6.12}$$

and

$$\tan\theta/\tan(px/c) = \frac{1+m\exp(-2\alpha x)}{1-m\exp(-2\alpha x)}. \tag{6.13}$$

Thus except when αx is very large the amplitude and phase vary with distance in a rather complicated manner and, although c can be determined comparatively easily from a set of phase measurements, the determination of α involves rather heavy computation.

The radar techniques developed during the Second World War have recently been applied to the measurement of elastic constants of solids. The method used has been to produce a short pulse of high-frequency oscillations and measure its time of transit and its attenuation as it passes across the specimen. This method has been used by Roth (1948), Mason and McSkimin (1947 and 1948), and by Ivey, Mrowca, and Guth (1949). The type of arrangement generally employed for this work is shown diagrammatically in Fig. 33. The pulses used were of between 1 and 15 microseconds duration and the oscillations of which they were composed have been of frequencies as high as 100 megacycles per second. The pulse travels back and forth along the specimen and is detected by the receiving crystal; the relative amplitudes of successive reflected pulses gives a measure of the attenuation, whilst the intervals between them give the velocity of propagation. Mason and McSkimin (1947 and 1948) and Roth (1948) have investigated metal specimens by this method, whilst Ivey, Mrowca, and Guth (1949) worked with rubber specimens. In all these experiments the ultrasonic pulses were communicated from the crystal to the specimen by means of an intermediate medium which in the experiments of Roth

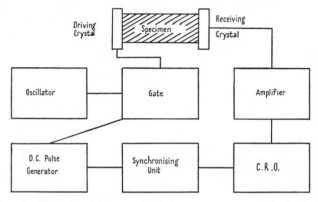

FIG. 33. Experimental arrangement for measurements with ultrasonic pulses.

and of Ivey, Mrowca, and Guth was a water bath, whilst Mason and McSkimin used wax joints for longitudinal pulses and layers of very viscous liquids for high-frequency distortional pulses. (Mason (1947) has shown that viscous liquids exhibit shear elasticity; polyisobutylene for example has a shear elasticity of 3×10^7 dynes/sq. cm. at frequencies of the order of 100 kc./sec., and such liquids can therefore be used for transmitting distortional pulses.)

The advantages of the pulse method are that once the apparatus has been set up the measurements of velocity and attenuation are rapid and simple and very high frequencies can be employed. Also with specimens of non-dispersive materials the method is capable of giving a very high degree of accuracy. The principal difficulty of the method, as pointed out by Davies (1950), often lies in the interpretation of the experimental results. When the lateral dimensions of the specimen are large compared with the length of the specimen and with the wavelength of the ultrasonic waves the time of transit will correspond to either the dilatational or the distortional wave velocity, depending on the type of pulse. When the specimens are in the form of rods, however, the pulses are reflected from the sides and a number of separate pulses arrive at the detector which have travelled by different routes, since, as shown in Chapter II, when a dilatational wave is incident obliquely on a free surface both a reflected dilatation wave and a reflected distortion wave are generated. Hughes, Pondrom, and Mims (1949) have carried out experiments on the transmission of pulses in metal rods and have shown that such a series of pulses is in fact obtained. Mason and McSkimin (1948) also have found that reflections from the sides of the specimen complicate the results when longitudinal pulses are used, but state that distortional waves travel without dispersion under these conditions since they are incident on the free surface at angles greater than the critical angle and are therefore reflected without change in form, the specimen acting as a wave guide. The propagation of continuous waves in very long rods was discussed in Chapter III, and it was shown that the velocity of propagation approaches the velocity of Rayleigh surface waves as the wavelength becomes short compared with the lateral dimensions of the rod (see Figs. 14 and 15).

The above considerations apply both to perfectly elastic and to dispersive media. When the medium is dispersive however, i.e. when the elastic properties vary with frequency, the interpretation of the results becomes more uncertain since there is no longer a unique velocity of propagation and the velocity at which energy is transferred is the group velocity c_g, which differs from the phase velocity c_p by an amount $\Lambda(dc_p/d\Lambda)$, where Λ is the wavelength. When the dispersion of a medium is high, as with some high polymers, this difference may be quite considerable. It is also necessary under these conditions for the pulse to consist of a large number of sinusoidal waves, since its Fourier spectrum will otherwise contain a wide band of frequencies which will be travelling with different velocities and the length of the pulse will increase as it progresses down the specimen. The variation in attenuation with frequency will further complicate matters as, in general, the high-frequency components will be damped more rapidly than the low-frequency ones and will be travelling faster so that they will appear at the front of the pulse.

From the above considerations it may be seen that the results obtained by the pulse method must be interpreted with some care, especially when the specimen is dispersive, but the method nevertheless provides a valuable means of studying the dynamic elastic properties of solids at high frequencies, and many useful results have already been obtained by it.

In the methods described so far the detection of the stress waves has depended on generating a small electric current which could then be amplified and measured. When, however, the material under investigation is transparent, there are a number of optical techniques which provide convenient methods of measuring the mechanical properties of materials without the use of so much auxiliary electronic equipment. One of these depends on the fact that when a transparent specimen is vibrating at very high frequency the repeating differences in density will result in its behaving like an optical diffraction grating, and if a beam of monochromatic light is passed through it a series of diffracted beams appear. From the diffraction patterns observed, the unit grating distance and hence the wavelength of the stress waves may be calculated and if the frequency of the oscillations is known the velocity of propagation may be determined. This method was first used by Debye and Sears (1932) and by Lucas and Biquard (1932) and has been applied by Schaefer and Bergmann (1934), Hiedemann, Asbach, and Hoesch (1934), and Hiedemann and Hoesch (1935) to the measurement of the elastic constants of transparent solids. An account of this work and of other optical diffraction methods of observing the propagation of ultrasonic waves is given in the book on ultrasonics by Bergmann (1949).

When the amplitude of the stress wave is sufficiently large the photo-elastic properties of a transparent solid may be utilized to observe the passage of stress waves. This technique depends on the fact that when most transparent solids are stressed they cease to be optically isotropic and become birefringent, i.e. the value of the refractive index depends on the plane of polarization of the incident light. If a specimen in the form of a plate is stressed it is found that at every point of the specimen there are two mutually perpendicular directions of polarization parallel to the plate for which the refractive index has a maximum and minimum value respectively, and these two directions correspond to directions in which the normal stress components at the point have maximum and minimum values. (These are the directions of *principal* stress and it may be shown that the shear stresses are zero along them.) For most solids it is further found that up to the elastic limit the difference between the extreme values of the refractive index is proportional to the algebraic difference in the values of the principal stresses. This result is known as *Brewster's law* and the constant of proportionality between the stress and the difference in refractive index is known as the *stress-optical coefficient* and is a physical constant of the material.

When a beam of plane polarized light passes through a stressed specimen it may be considered as resolved into two components which are plane polarized in directions perpendicular to each other and parallel to the directions of principal stress. These two beams travel through the sample with different velocities so that a phase difference is introduced between them. In general the light which emerges is elliptically polarized and if this emergent light is viewed through an analyser, e.g. a nicol prism or a sheet of polaroid, the observed intensity depends on the phase difference introduced by the specimen and hence on the applied stress. When monochromatic light is used the specimen will appear to be traversed by a number of light and dark fringes, the pattern depending on the stress distribution in the specimen, and from such stress patterns the difference in the principal stresses throughout the specimen may be determined. The photo-elastic method has been widely used to determine experimentally the static stress distribution in various types of engineering structure and accounts of the theory and the application of the method to engineering problems will be found in the books on the subject by Coker and Filon (1931), Frocht (1941), and Jessop and Harris (1949).

To apply the photo-elastic technique to the study of the propagation of stress waves it is necessary to take high-speed photographs of the photo-elastic patterns, and this has been done using spark discharges by Schardin and Struth (1938), Senior and Wells (1946), Schardin (1950), and Christie (1952). One set of such photographs taken by Mr. D. G. Christie is shown in Plate I (frontispiece). In these photographs the propagation of a stress pulse in a sheet of 'Perspex' is shown; the stress pulse has here been produced by a small charge of lead azide which was detonated in contact with the upper edge of the sheet. (In this experiment circularly polarized light was used, since with plane polarized light *isoclinics* appear. These are dark bands which correspond to regions where the principal stresses are parallel to the axes of polaroids, and their presence tends to confuse the pattern.)

This set of photographs shows the propagation of two types of cylindrical wave which travel with different velocities from the seat of the charge. The faster wave is a longitudinal wave which, so long as the wavelength is large compared with the thickness of the plate, travels with a velocity $[E/\rho(1-\nu^2)]^{\frac{1}{2}}$ (see equation (3.91), Chapter III), whilst the slower wave is a transverse wave which appears to travel with the distortional velocity in the material, $[\mu/\rho]^{\frac{1}{2}}$. This transverse wave is a result of the distortion of the top edge of the plate caused by the charge, and the particle motion is parallel to the plane of the plate. These transverse waves do not appear when a charge is detonated at the centre of a plate. In the later pictures the reflection of the stress waves at the sides and bottom of the plate takes place, and the superposition of the incident and reflected waves may be seen to produce very complex stress patterns. The time intervals between the sparks producing the separate photographs were measured with a photocell and cathode-ray oscillograph and are accurate to 0·1 microseconds, so that this technique affords a fairly accurate method of measuring the velocity of propagation of stress waves in transparent materials and also provides a means of studying the reflection of stress waves at interfaces. In its application to the measurement of dynamic elastic constants, however, it suffers from the disadvantage inherent in all pulse methods, namely that with dispersive systems the interpretation of the results is very difficult since the pulse changes shape as it travels through the medium and there is no unique velocity of propagation. Further, it is difficult to produce a pulse of sufficient amplitude except by the use of explosives or projectiles and these pulses, unlike the wave packets from piezo-electric crystals will contain a wide spectrum of Fourier components and will consequently be dispersed very rapidly.

Dynamic stress-strain measurements

In the three methods of measuring dynamic elastic properties of solids which have been discussed, viz.: free oscillations, forced oscillations, and wave propagation, the elastic constants and the internal friction could not be derived from the measurements unless certain assumptions were made about the nature of the dissipative forces and the linearity of the system. The assumptions made were that the dissipative force was proportional to the time rate of change of the strain and that the type of mechanical behaviour was independent of the amplitude of the deformation within the range of stress employed in the experiments. Assuming Boltzmann's superposition principle to hold, the *memory* function could then be obtained from a series of measurements made over a range of frequencies and from this the mechanical behaviour of the solid when subjected to a non-harmonic stress could in theory be derived.

It is often preferable, however, to make direct measurements of stress and strain whilst the specimen is being deformed, since the mechanical behaviour for a given stress cycle can then be obtained without making any *a priori* assumptions about the behaviour of the solid, and by this method the shape of the hysteresis loop is obtained instead of merely the amount of the energy lost in a stress cycle. At high rates of loading, however, the experimental determination of the stress-strain curve involves very considerable difficulties. These are associated with inertia effects in the measuring apparatus and with the recording of transient stresses and strains.

As the rate of loading is increased the acceleration of any moving parts of the straining apparatus begins to require forces comparable with those necessary to deform the specimen. It is often impossible to separate these inertia effects from the effects due to the mechanical properties of the specimen under investigation, and at the highest rates of loading the inertia of the specimen itself will result in a non-uniform distribution of stress along its length, the problem having become one of wave propagation through the specimen with repeated reflections from the ends. At such high rates of loading, the use of ordinary mechanical stress gauges is precluded since their natural period is far too long for them to give significant readings when the forces are changing rapidly.

The problems involved in such measurements and some of the methods which have been used to overcome them are discussed by Taylor (1946). Much of the work on the dynamic mechanical testing of solids has been concerned with measurements of the yield point and tensile strength at high rates of loading; this is discussed in Chapter VIII. We shall here describe the methods of measuring stress-strain curves at high rates of loading which have been developed by Taylor (1946), E. Volterra (1948), and by the author (Kolsky, 1949).

Taylor and E. Volterra used a photographic method of recording the stress and strain in their specimens, which were in the form of short cylinders. These were placed on the plane end of a cylindrical bar which hung as a ballistic pendulum. Another freely suspended bar was swung against it so that on impact the specimen was compressed between the flat end faces of the two bars. The strain-time relation was derived directly from the photographic record, whilst the stress-time relation was obtained from the movement of the steel bar, which was initially at rest and which was accelerated by the stress communicated to it by the specimen. Obtaining the stress-time curve thus involves a double differentiation of the displacement-time curve, but the measurements from the photographic records were found to be sufficiently accurate for this to be done. Specimens of rubber and other high polymers were investigated by this method, the duration of the stress cycle being between 5 and 17 milliseconds, and stress-strain curves were obtained. In order to analyse the results it was assumed that the materials obeyed Boltzmann's superposition principle and the form of relation between the stress σ and the strain ϵ employed was

$$\sigma = \psi(\epsilon) + \int\limits_0^t \phi(t-T)\frac{d\epsilon}{dT}\,dT. \qquad (6.14)$$

On comparing this with equation (5.38) it may be seen that the relation is here of a somewhat different form. The method of deriving (6.14) is to consider the small changes in the strain which have occurred in the past history of the specimen and sum the residual elements of stress produced by them. Thus $\phi(t-T)\,d\epsilon$ is the stress remaining at time t due to a change in strain of $d\epsilon$ which took place at time T. Gross (1947) has compared different forms of the superposition principle and has shown that they are equivalent and that the relaxation time spectrum can be derived from either form of memory function.

Taylor and Volterra have interpreted the results of the ballistic pendulum experiments on polythene, a wax-like plastic, in terms of (6.14) and showed that their experimental results could be fairly represented by taking the memory function $\phi(t-T)$ to be of the form $A\exp[(T-t)/\tau]$, with $A = 2\times 10^9$ and $\tau = 1/600$ in C.G.S. units. This is equivalent to assuming that the material behaves as if it had a single relaxation time of $1\cdot7$ milliseconds. It should be remembered that the times of impact in Taylor and Volterra's experiments were between 5 and 17 milliseconds, so that the relaxation time obtained is about one-tenth of the longest time observed. As is shown later in this section experiments carried out with stress cycles the duration of which was of the order of 20 microseconds gave a value of τ of 2 microseconds, so that the use of a single relaxation time for this material appears to have little theoretical significance and is only useful as a method of summarizing its mechanical behaviour for stresses which are applied for a limited range of times. This point will be further discussed later in this chapter.

Volterra (1948) also describes the use of the Davies pressure bar, which was discussed in Chapter IV, for measuring the stress-strain relation of materials in the form of cylinders. One face of the specimen was placed against the firing-end of the bar, and an anvil in the form of a short steel cylinder was placed in contact with the opposite face of the specimen. A 0·22-inch bullet was then fired at the exposed face of the anvil. The stress pulse which was produced at the end of the bar by the pressure of the specimen travelled down the bar and was recorded by the use of a parallel-plate condenser microphone and cathode-ray oscillograph as in Davies's apparatus (see Fig. 23). Thus the stress-time curve for the specimen could be found. In order to obtain the strain-time curve a similar bullet was fired at the end of the Davies bar with no anvil or specimen present and the pressure-time curve so obtained was assumed to be similar to that exerted on the anvil in the first experiment. If the mass of the anvil is M and its displacement at time t is x we have from Newton's second law

$$M\frac{d^2x}{dt^2} = P_1(t) - P_2(t), \qquad (6.15)$$

where $P_1(t)$ is the force exerted by the bullet on the anvil and $P_2(t)$ is the force between the specimen and the anvil.

Now if the displacement of the end of the pressure bar as measured by the condenser microphone is $\xi_1(t)$ when the specimen is present and $\xi_2(t)$ when there is no specimen, we have

$$P_1(t) = \tfrac{1}{2}B\rho c\frac{d\xi_1}{dt} \quad \text{and} \quad P_2(t) = \tfrac{1}{2}B\rho c\frac{d\xi_2}{dt}, \qquad (6.16)$$

where B is the cross-sectional area of the specimen, ρ is the density of the material of the bar, and c is the velocity of propagation of longitudinal waves in it (see equation (4.1)). Thus from (6.15) we have on integrating twice and substituting from (6.16)

$$x = \frac{B\rho c}{2M}\int_0^t (\xi_1 - \xi_2)\, dt. \qquad (6.17)$$

From (6.17) the strain-time curve can be obtained, and hence the stress-strain curve for the specimen can be derived.

The main disadvantage of the method is that the stress-strain curve can be obtained only whilst the stress is increasing, and that the pressure pulses must be sufficiently long for them not to be distorted in their passage down the steel bar. By this method

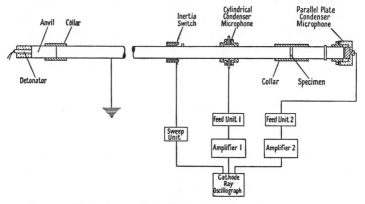

Fig. 34. Davies bar used for dynamic stress-strain measurements.

Volterra (1948) made measurements with copper and polythene specimens. With the copper specimens the stress reached its maximum in 130 microseconds, whilst with polythene specimens it took 320 microseconds to do so.

Fig. 34 shows a modified form of the Davies bar which has been used by the author (Kolsky, 1949) to measure the stress-strain behaviour of disk-shaped specimens when they are taken round stress-cycles in times of the order of 20 microseconds. The pressure pulse is here produced by a detonator which is held against a replaceable steel anvil at the firing end of the bar. The pulse travels down the bar and compresses the specimen which is placed between the end faces of the main bar and an extension bar, a closely fitting collar holding the specimen and extension bar in position. The flat surfaces of the specimens are lubricated with a thin layer of grease to allow free lateral movement. The amplitude of the pulse before reaching the specimen is measured with a cylindrical condenser microphone, the output of which is amplified and fed in to one set of Y deflexion plates of a double-beam cathode-ray oscillograph. The output from a parallel-plate condenser microphone at the end of the extension bar is also amplified and fed on to the other set of Y plates of the oscillograph. An inertia switch similar to that described by Davies is used to trigger the sweep unit which gives the time sweep in the X-direction in the oscillograph and brightens the oscillograph spot, so that a photographic record of the trace can be obtained. The oscillograph trace obtained from the cylindrical microphone gives the pressure-time relation for the incident pulse travelling up the bar to the specimen (see equation (4.2)); we shall call this stress $\sigma_1(t)$. The trace for the parallel-plate condenser microphone gives the displacement of the free end of the extension bar, and we will call this $\xi_2(t)$. From $\sigma_1(t)$ and $\xi_2(t)$ the stress-strain curve of the specimen may be derived.

The stress-time relation $\sigma_2(t)$ may be obtained from the equation

$$\sigma_2(t) = \tfrac{1}{2}\rho c \frac{d\xi_2}{dt}. \tag{6.18}$$

In order to find the strain-time relation we must consider the displacements of the two ends of the specimen separately. The end of the specimen in contact with the main bar is displaced by both the incident pulse and by the pulse which is reflected back along the bar. The displacement due to the incident pulse is given by $(1/\rho c) \int_0^t \sigma_1(t)\, dt$, whilst the displacement due to the reflected pulse is $(1/\rho c) \int_0^t [\sigma_1(t) - \sigma_2(t)]\, dt$. Thus the total displacement of this face is

$$\frac{1}{\rho c} \int [2\sigma_1(t) - \sigma_2(t)]\, dt.$$

The displacement of the face in contact with the extension bar is $(1/\rho c) \int_0^t \sigma_2(t)\, dt$ so that the change in length of the specimen is

$(2/\rho c) \int\limits_0^t [\sigma_1(t) - \sigma_2(t)] \, dt$, and from (6.18) this equals

$$\frac{2}{\rho c} \int\limits_0^t \sigma_1(t) \, dt - \xi_2(t)$$

$$= \xi_1(t) - \xi_2(t) \quad \text{if we write } \xi_1(t) \text{ for } \frac{2}{\rho c} \int\limits_0^t \sigma_1(t) \, dt.$$

It may be seen that $\xi_1(t)$ is the displacement of the end of the extension bar when there is no specimen present between the bars and the initial pulse travels into the extension bar without change in form. (As pointed out by Davies (1948) the distortion observed with a cylindrical condenser microphone with short pulses is greater than that observed with a parallel-plate instrument and in the experiments described here $\xi_1(t)$ was measured directly by separate experiments with no specimen between the bars.)

In these experiments the length of the specimen had to be sufficiently short compared with the length of the pulse for the difference in pressure on both its faces to be neglected. Under these conditions the forward acceleration of the specimen could be neglected, but allowance had still to be made for the radial motion since there will be kinetic energy associated with this radial motion however thin the specimen may be. This radial energy correction involves the second differential coefficient of the strain with respect to time. Fortunately in the conditions of the experiment the correction never amounted to more than a few per cent. and its absolute value did not therefore have to be determined with any great accuracy.

The stress-strain curves for several plastics and rubbers were measured by this method, and some experiments were also carried out with copper and lead specimens. The results for the plastics and rubbers were interpreted in terms of a memory function using the form of function given by equation (6.14). In the case of polythene it was found that fair agreement with the experimental results could be obtained by assuming that the function $\phi(t-T)$ was of the form $A \exp[-(t-T)/\tau]$ with

$A = 4.6 \times 10^{10}$ dynes/sq. cm. and $\tau = 2$ microseconds. On comparing these values with those obtained by Volterra it may be seen that the values of A and τ are here very different from those derived from Volterra's experiments, and this shows that the assumption of a single relaxation time is an over-simplifica-

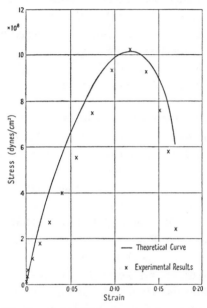

Fig. 35. Comparison between experimental results for polythene and curve obtained from the expression

$$\sigma = 10^9\epsilon + 4.6 \times 10^{10} \int_0^t \exp[-(t-T) \times 5 \times 10^5] \frac{d\epsilon}{dT} \, dT.$$

tion of the problem. Over a short range of time the very simple form of memory function is quite adequate, however, as can be seen from Fig. 35 in which the experimental points for polythene are compared with the theoretical curve. It may be seen that reasonable agreement has been obtained. The experimental points are at 2-microsecond intervals so that the complete stress cycle may be seen to correspond to about 28 microseconds.

As mentioned in Chapter V the general form of the memory function corresponds to a spectrum of relaxation times and the reason why a single relaxation time gives good agreement when only a limited time and range is considered is that processes which are associated with times very long compared with the time of application of the stress will not influence the mechanical behaviour, whilst processes with very short relaxation times will appear as part of the 'instantaneous' mechanical response of the material and not as part of the memory function. (The memory function $\phi(t-T)$ is found to decrease monotonically, and mathematically the use of a single relaxation time is equivalent to finding an exponential which will roughly coincide with it over a small range of values of $(t-T)$.)

Discussion of experimental results

Most of the experimental work on the dynamic elastic behaviour of solids has been carried out on specimens either of metals or of high polymers. There have, however, been a number of papers describing work with other materials and recent investigations include measurements in the audiofrequency range on wood by Kruger and Rohloff (1938), Barducci and Pasqualini (1948), and Fukada (1950). At ultrasonic frequencies measurements on various glasses have been made by Allegretti (1948), whilst Hunter and Siegel (1942) have studied the elastic constants and internal friction of rock salt near its melting-point. The elastic constants of alkali halide crystals have also been measured by Nurmi (1941), Huntingdon (1947), and Galt (1948). Thiede (1941) describes some measurements on the German material *Degussit* (aluminium oxide). This solid is remarkable for its exceptionally high velocity of sound, 9,600 metres per second.

The work on metals in recent years has been largely concerned either with the exact determinations of elastic constants or with the study of the various mechanisms which result in internal friction. These latter investigations were discussed at the end of Chapter V, and as was shown there the observed results are found to be in good agreement with the theoretical predictions made by Zener of the energy losses due to thermal conduction. Zener (1949) gives a general review of losses due to internal friction in metals and discusses the type of relaxation time spectrum which is observed.

The theoretical interpretation of the work on high polymers has been less successful, as a large number of different molecular mechanisms would appear to be involved when a high polymer is deformed. The relaxation spectra of such materials tend to be very flat, and measurements over several decades of frequency are required in order to show any general trends. The main conclusion so far drawn from the experimental results is the marked dependence of the mechanical properties of these materials on temperature. The effect of increasing the temperature is found to be equivalent to that of lowering the frequency, and vice versa. Alexandrov and Lazurkin (1940) were the first to carry out a full investigation of the effect of temperature on the dynamic elastic properties of rubber. They worked at frequencies between 0·1 and 1,000 cycles per minute and at temperatures between −180° C. and 200° C. Treloar (1949) discusses their results and those of other workers on the dynamic properties of rubber.

Nolle (1949), whose experimental work has already been described, covered the largest frequency range with rubber-like materials at different temperatures. He obtained results for Young's modulus at frequencies between 0·1 cycles per second and 120 kilocycles per second for various rubbers. In Nolle's experiments the vibrations were either longitudinal or flexural so that Young's modulus was the relevant elastic constant. By analogy with electrical measurements Nolle expresses his results in terms of a complex Young's modulus of the form $E_1 + iE_2$. The real part of the modulus, E_1, corresponds to the elastic restoring force and for a perfectly elastic material is equal to Young's modulus, whilst E_2, the imaginary part, is a measure of the mechanical loss in the material. The basis of this notation is the assumption that at any one frequency the material may be considered as a Voigt solid for which the stress-strain relation will be of the form

$$\sigma = E\epsilon + E'\frac{d\epsilon}{dt} \quad \text{(see equation (5.29)).}$$

E is here Young's modulus and E' is the 'coefficient of normal viscosity'. Now if the material is performing sinusoidal vibrations of frequency $p/2\pi$ we have

$$\epsilon = \epsilon_0 \exp(ipt),$$

so that the relation between stress and strain may be written

$$\sigma = [E + ipE']\epsilon.$$

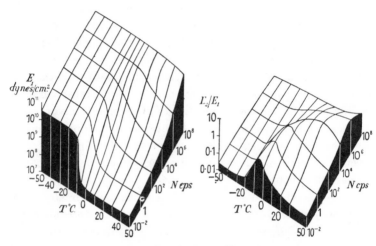

FIG. 36. Sketches showing approximate behaviour of dynamic modulus E_1 and loss factor E_2/E_1 for a Buna rubber (Nolle).

Thus the expression $(E + ipE')$ here takes the place of Young's modulus E for a perfectly elastic material. In Nolle's notation E_1 is written for E, and E_2 for pE'. As shown in Chapter V, (equation (5.31)), the logarithmic decrement is approximately equal to $p\pi E'/E$ so that in Nolle's notation it will be given by $\pi E_2/E_1$.

Fig. 36 is a schematic summary of the way in which E_1 and E_2/E_1 vary with temperature and frequency in a sample of Buna rubber. These curves are as given by Nolle, who states that they are only semi-quantitative. The experimental results of Young's modulus did not extend beyond 10^5 cycles per second, the approximate upper limits of E_1 and E_2/E_1 being deduced from some measurements of the velocity and attenuation of dilatational waves of frequency 15 megacycles per second.

The upper curve shows that E_1 increases with frequency and decreases with temperature. The freezing-point of the material, which corresponds to the large rise of modulus, occurs at about $-20°$ C. at frequencies below one cycle per second but at higher temperatures when higher frequency oscillations are employed. Thus when subjected to rapidly varying stresses the material appears to 'freeze' at a higher temperature than when deformed more slowly. This effect was first pointed out by Alexandrov and Lazurkin (1940). The internal friction E_2/E_1 for any one frequency may be seen to have a maximum with respect to temperature, and the temperature at which this maximum loss occurs increases with frequency. The sketch given by Nolle also shows a decrease in E_2/E_1 at the very highest frequencies. There appears to be little direct evidence in these experiments that such a decrease in fact occurs, and the main reason Nolle had for assuming that it does so is that there is a close analogy in other respects between internal friction in high polymers and dielectric loss in polar substances. Since the dielectric loss is found to pass through a maximum with frequency at any one temperature the internal friction might also be expected to do so. Experiments on the propagation of high-frequency dilatational pulses through various rubbers which have been carried out by Ivey, Mrowca, and Guth (1949) do, however, clearly show that the internal friction passes through a maximum as the frequency is increased.

On comparing the curves for one temperature in Fig. 36 with the theoretical curves for a single relaxation time shown in Fig. 28 it may be seen that in both cases the damping loss shows a maximum whilst the effective elastic modulus (given by the velocity curve in Fig. 28) is S-shaped. The experimental curves for rubber are, however, very much flatter than the theoretical curve for a material with a single relaxation time, and the former may be considered as having been produced by the superposition of curves from a 'spectrum' of relaxation times. Nolle (1949) has evaluated an approximate relaxation time spectrum in terms of Maxwell elements, and Fig. 37 shows the function $A(\log_e \tau)$ plotted against frequency. The theory of a relaxation time spectrum was discussed in Chapter V and the relation between $A(\tau)$ and the Boltzmann memory function is given in equation (5.20). It may be seen from the figure that the

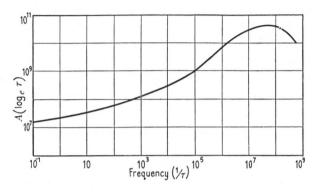

Fig. 37. Approximate relaxation time spectrum for a Buna rubber (Nolle).

relaxation time spectrum is very flat and it is difficult to come to any definite conclusions about the molecular processes which produce mechanical relaxation from it. The spectrum is, however, a convenient way of summarizing the results of the experiments over the very wide frequency range covered.

CHAPTER VII

PLASTIC WAVES AND SHOCK WAVES

IN the first part of this monograph the propagation of stress waves in perfectly elastic media was considered. The effect of dissipative forces which result in the conversion of elastic energy into heat has been discussed in Chapter V, and the experimental results of dynamic measurements were described in Chapter VI. The systems so far treated have, however, all obeyed linear differential equations and it has been assumed that the stress amplitude was in all cases sufficiently small for the elastic restoring force to be proportional to the strain. The general problem of the propagation of stress in a non-linear medium is clearly an extremely difficult one, but solutions have been obtained in a few special cases and these will be discussed in this chapter.

The problem we shall consider first is that of the propagation of plastic deformation along a wire or rod which is perfectly elastic for stresses up to a given value known as the *proportional limit*, but for stresses greater than this there is a different relation between stress and strain, and on the removal of the stress hysteresis is observed. Thus for stresses greater than the proportional limit the strain is a univalued function of the stress only so long as the stress is increasing; for decreasing stresses a different stress-strain relation applies. This problem was first considered during the Second World War independently by Taylor (1946), von Kármán and Duwez (1950),† and Rakhmatulin (1945).

In treating the problem, von Kármán and Rakhmatulin used Lagrangian coordinates, whilst Taylor used the coordinates of Euler. When suitable transformations of the notations are made the two treatments may be shown to give the same

† The work of Taylor and von Kármán first appeared as confidential government reports in 1940–2, and the references given here are to subsequent publications by these authors.

result, but as the physical approach is different in the two cases summaries of both treatments will be given here.

The system considered is that of an infinitely long wire or rod the end of which is suddenly given a velocity V_1 at time $t = 0$, this velocity being maintained constant for $t > 0$. The problem is to determine the conditions at time t. It is assumed that the wire has a linear stress-strain relation up to a critical value of the stress, and that above this the relation between stress σ and strain ϵ is no longer linear but is univalued for increasing stresses, so that $\sigma = \sigma(\epsilon)$, where $\sigma(\epsilon)$ is a known function of strain ϵ. It is also assumed that the stress-strain relation is independent of the rate of loading and that the cross-section of the wire is sufficiently small for the effects of radial kinetic energy to be ignored.

Plastic waves by the Lagrangian method

We here fix our attention on a small segment of wire which at time $t = 0$ was a distance X from the end. The initial position of the impelled end of the wire is taken as the origin, the positive direction being along the wire. If at time t the displacement of the segment is u, we have from Newton's second law of motion

$$\rho_0 \frac{\partial^2 u}{\partial t^2} = \frac{\partial \sigma'}{\partial X}, \qquad (7.1)$$

where ρ_0 is the density of the *unstretched* wire and σ' is the engineering stress, i.e. the force acting on the wire divided by its *original* cross-sectional area.

Now if it is assumed that the relation between σ' and the strain is univalued when the stress is increasing we may write for (7.1)

$$\rho_0 \frac{\partial^2 u}{\partial t^2} = \frac{d\sigma'}{d\epsilon} \frac{\partial \epsilon}{\partial X}. \qquad (7.2)$$

ϵ is here the strain and is defined as the ratio between the change in length of an element to its original length, thus $\epsilon = \partial u/\partial X$. It is not assumed that ϵ is necessarily small. Now $d\sigma'/d\epsilon$ is the modulus of deformation, elastic or plastic, and we will call this quantity S, S being considered as a *known* function $S(\epsilon)$ of ϵ. (7.2) then becomes

$$\rho_0 \frac{\partial^2 u}{\partial t^2} = S \frac{\partial^2 u}{\partial X^2}. \qquad (7.3)$$

The boundary conditions are that $u = V_1 t$ at $X = 0$ for all positive values of t, and $u = 0$ at $X = \infty$. The problem is to solve (7.3) for a material for which the stress-strain curve, and hence the function $S(\epsilon)$, is known. There are two solutions of (7.3) as it may be seen on inspection that

$$u = V_1 t + \epsilon_1 X \qquad (7.4)$$

will satisfy (7.3) and the boundary condition at $X = 0$, and this solution corresponds to a constant strain ϵ_1.

Another solution can be found when S/ρ_0 is equal to X^2/t^2. Since S is a function of ϵ this solution means that ϵ is a function of the variable (X/t), which we shall call β. Now assume $\epsilon = f(\beta)$, then

$$u = \int_\infty^X \frac{\partial u}{\partial X}\, dX = \int_\infty^X f(\beta)\, dX = t \int_\infty^\beta f(\beta)\, d\beta \qquad (7.5)$$

(since $dX = t\, d\beta$). Differentiating (7.5) twice with respect to t we have

$$\frac{\partial^2 u}{\partial t^2} = \frac{\beta^2}{t} f'(\beta), \qquad (7.6)$$

where $f'(\beta)$ is the derivative of $f(\beta)$ with respect to β and

$$\frac{\partial^2 u}{\partial X^2} = \frac{1}{t} f'(\beta). \qquad (7.7)$$

Substituting from (7.6) and (7.7) in (7.3) we find that either

$$f'(\beta) = 0 \qquad (7.8)$$

or
$$\rho_0 \beta^2 = S. \qquad (7.9)$$

(7.8) corresponds to the solution given by (7.4), whilst (7.9) corresponds to a solution in which (X/t) is equal to $(S/\rho_0)^{\frac{1}{2}}$.

The complete solution is that there are three regions in the wire:

(a) from $X = 0$ to $X = Ct$ the strain is a constant ϵ_1, where C is the velocity of propagation of the plastic wave front;

(b) between $X = Ct$ and $X = c_0 t$ the relation $X^2/t^2 = S/\rho$ holds (c_0 is here the velocity of propagation of the elastic wave);

(c) for $X > c_0 t$, $\epsilon = 0$.

(At $X = c_0 t$ there is a discontinuity in the strain. S is equal to Young's modulus E and we have $c_0^2 = E/\rho$ as in the elastic case.)

Fig. 38 shows the relation between ϵ and β schematically, the three regions corresponding to

$$(a) \quad 0 < \beta < C,$$
$$(b) \quad C < \beta < c_0,$$
$$(c) \quad \beta > c_0.$$

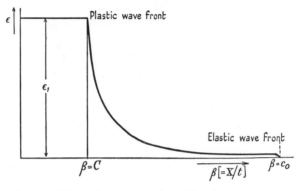

FIG. 38. Schematic representation of distribution of strain.

In the region (b) between the elastic wave front and the plastic wave front the strain is variable, since every strain increase from ϵ to $\epsilon + d\epsilon$ is propagated with a velocity which depends on the value of S at that particular value of ϵ. To complete the solution we must find the relation between C, the velocity of the plastic wave front, and V_1, the velocity with which the end is moving.

Now from equation (7.5) we have for the displacement of the end of the wire

$$u(0, t) = V_1 t = t \int_{\infty}^{0} f(\beta)\, d\beta,$$

or

$$V_1 = \int_{\infty}^{0} f(\beta)\, d\beta. \tag{7.10}$$

And from Fig. 38 it may be seen that the integral which is the area under the curve may be expressed as

$$V_1 = -\int_{0}^{\epsilon_1} \beta\, d\epsilon = -\int_{0}^{\epsilon_1} \left(\frac{S}{\rho_0}\right)^{\frac{1}{2}} d\epsilon. \tag{7.11}$$

Thus if S is known as a function of ϵ (7.11) gives a relation between ϵ_1, the strain behind the plastic wave front, and V_1, the velocity with which the end of the wire is being stretched. The velocity of propagation of the plastic wave front C can then be obtained from the stress-strain curve since it is equal to the value of $(S/\rho_0)^{\frac{1}{2}}$ at ϵ_1.

The distribution of stress σ' in the wire can be deduced from the distribution of strain ϵ, and there will be a certain value of the strain ϵ_1 corresponding to the tensile strength of the wire. The velocity V_1 corresponding to this value of ϵ can be found from (7.11), and it would be expected that the wire will break instantly if it is stretched at velocities above this critical value.

Plastic waves by the Eulerian method

In the treatment of the problem given above, the equations were derived for a small section of wire which was moving in space and changing its length and cross-sectional area as the wire was stretched. When the problem is treated by the method of Euler, as was done by Taylor, we fix our attention on a fixed region of space and obtain the equations of motion and continuity for the wire which enters and leaves this region.

If we now take our x-coordinate to be the distance from the initial position of the end of the wire, with x increasing along the wire, the equation of motion for the region with coordinates between x and $x+\delta x$ may be obtained by equating the difference in the forces acting on the two sides of the element of wire present there at time t to the product of the mass and the acceleration of this element. Let the engineering stress at x be σ' as in the previous section, then at $x+\delta x$ it will be

$$\sigma' + \frac{\partial \sigma'}{\partial x}\, \delta x,$$

so that if the original cross-sectional area of the wire is A_0 the net force acting is $A_0(\partial\sigma'/\partial x)\,\delta x$. The mass of the element is $A_0\rho_0\,\delta x/(1+\epsilon)$, where ρ_0 is the density of the material of the wire before deformation and, as before, ϵ is the strain. If the velocity of the element is V at time t its centre will have moved a distance

$V \, \delta t$ in time δt, so that its velocity at time $t + \delta t$ will be

$$V + V \, \delta t \, \frac{\partial V}{\partial x} + \frac{\partial V}{\partial t} \, \delta t.$$

The acceleration is therefore $\partial V / \partial t + V \partial V / \partial x$. From Newton's second law of motion we thus obtain

$$\frac{\rho_0 A_0}{1 + \epsilon} \left[\frac{\partial V}{\partial t} + V \frac{\partial V}{\partial x} \right] \delta x = A_0 \frac{\partial \sigma'}{\partial x} \, \delta x$$

or
$$\frac{\rho_0}{1 + \epsilon} \left[\frac{\partial V}{\partial t} + V \frac{\partial V}{\partial x} \right] = \frac{\partial \sigma'}{\partial x} = \frac{d \sigma'}{d \epsilon} \frac{\partial \epsilon}{\partial x}. \tag{7.12}$$

If the mass per unit length of the wire at any time is M, we have

$$M = \frac{\rho_0 A_0}{1 + \epsilon}. \tag{7.13}$$

We may equate the time rate of change of this linear density to the difference between the mass entering and the mass leaving an element of space. This gives

$$\frac{\partial M}{\partial t} = - \frac{\partial}{\partial x} (M V). \tag{7.14}$$

Substituting from (7.13) for M we finally obtain

$$\frac{\partial \epsilon}{\partial t} + V \frac{\partial \epsilon}{\partial x} = (1 + \epsilon) \frac{\partial V}{\partial x}, \tag{7.15}$$

which is the equation of continuity for the wire.

Now (7.12) and (7.15) are similar to the equations for the propagation of a plane wave of finite amplitude in a compressible fluid (see Lamb, 1932, pp. 481–4), and this problem has been treated by Earnshaw and Riemann. Following Earnshaw's method of solution we assume that V is a function of the strain only, so that we may write

$$V = f(\epsilon). \tag{7.16}$$

(7.12) then becomes

$$\frac{\rho_0}{1 + \epsilon} \frac{df}{d \epsilon} \left[\frac{\partial \epsilon}{\partial t} + V \frac{\partial \epsilon}{\partial x} \right] = \frac{d \sigma'}{d \epsilon} \frac{\partial \epsilon}{\partial x}, \tag{7.17}$$

and (7.15) gives

$$\frac{\partial \epsilon}{\partial t} + V \frac{\partial \epsilon}{\partial x} = \frac{df}{d\epsilon} \frac{\partial \epsilon}{\partial x} (1+\epsilon). \tag{7.18}$$

Substituting for $\partial \epsilon / \partial t + V \partial \epsilon / \partial x$ from (7.18) in (7.17) we obtain

$$\left(\frac{df}{d\epsilon}\right)^2 = \frac{1}{\rho_0} \frac{d\sigma'}{d\epsilon}. \tag{7.19}$$

To obtain the velocity $V = f(\epsilon)$ we must integrate (7.19), and if we do this from the end of the wire which is at rest and where $\epsilon = 0$, we have

$$V = - \int_0^\epsilon \left[\frac{1}{\rho_0} \frac{d\sigma'}{d\epsilon}\right]^{\frac{1}{2}} d\epsilon. \tag{7.20}$$

This is identical with the result for the velocity of the end of the wire obtained by the Lagrangian method and given in equation (7.11), since $S = d\sigma'/d\epsilon$.

Substituting for $df/d\epsilon$ in equation (7.18) from (7.19) we have

$$\frac{\partial \epsilon}{\partial t} + \left[V + (1+\epsilon)\left(\frac{S}{\rho_0}\right)^{\frac{1}{2}}\right] \frac{\partial \epsilon}{\partial x} = 0$$

or, writing c for $(1+\epsilon)(S/\rho_0)^{\frac{1}{2}}$,

$$\frac{\partial \epsilon}{\partial t} + (c+V) \frac{\partial \epsilon}{\partial x} = 0. \tag{7.21}$$

Thus for points in the wire which move in space with velocity $(c+V)$ the strain ϵ and hence the stress σ' are constant. It should be noted that since x is measured up the wire V will be negative for stretching.

For most solids, as mentioned in the previous section, $d\sigma'/d\epsilon$ is constant for stresses up to the elastic limit and then decreases, so that large strains are propagated with velocities lower than that of elastic waves and the distribution of strain will be similar to that shown in Fig. 38. When, however, $d\sigma'/d\epsilon$ increases with increasing ϵ, large strains will travel faster than small ones and any large pulse travelling through the medium acquires a steep front, the gradient of which is ultimately limited by dissipative forces such as viscosity and thermal conduction. These dissipative forces become increasingly important as the pressure gradient at the front of the pulse becomes steeper.

The formation of such *shock waves* will be discussed later in this chapter.

Waves of unloading

We have so far considered what occurs in a specimen of infinite length when one end of it is stretched with constant velocity. As mentioned earlier, a different stress-strain relation

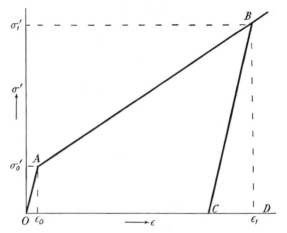

Fig. 39. Idealized stress-strain curve of plastic material.

applies when the stress is decreasing. In general the propagation of the wave of unloading produced when the end is released, and the interference between this wave with those already travelling in the specimen is a very complex problem. For purely elastic deformations the problem has been investigated by Perry (1906) in his paper on 'Winding Ropes in Mines', and we shall here consider the behaviour of a plastic material with an idealized stress-strain relation as shown in Fig. 39. The stress-strain curve is here assumed to be linear and reversible up to the point A, the proportional limit, and at this point to decrease suddenly in gradient but remain linear. Further, if the stress is reduced at any point B after the proportional limit has been passed, the curve is assumed to return along BC, which is parallel to OA, and a residual strain OC remains when the stress has been removed.

The specimen is assumed then to have become elastic for stresses up to the value at B, the curve BC being reversible.

If the end of a wire of material having such a stress-strain curve is stretched with constant velocity a steep-fronted plastic wave is obtained and the distribution of stress, strain, and particle velocity for an infinite wire is as shown in Fig. 40 (a). If the modulus $d\sigma'/d\epsilon$ is E in the elastic region (the gradient of OA in Fig. 39) and its value for increasing stresses in the plastic region is S_1 (the gradient of AB), the velocity of propagation of the elastic wave relative to the wire is $c_0/(1+\epsilon)$, where $c_0 = (E/\rho)^{\frac{1}{2}}$, and the corresponding velocity of the plastic wave is $C/(1+\epsilon)$, where $C = (S_1/\rho)^{\frac{1}{2}}$. Between the elastic and plastic wave fronts the stress and the strain are σ_0 and ϵ_0, the values at the proportional limit A, and the particle velocity is $\sigma_0/(\rho c_0)$. Behind the plastic wave front the strain ϵ_1 is given by equation (7.11) and the stress σ_1 associated with it may be found from the stress-strain curve, Fig. 39. It may be seen to be equal here to $E\epsilon_0 + S_1(\epsilon_1 - \epsilon_0)$. The particle velocity in this region is equal to V_1, the velocity with which the end of the wire is being extended, the part of the wire in which plastic flow has occurred moving as a rigid body.

Now when the end of the wire is suddenly released so that the stress at the end of it drops to zero, a wave of unloading travels down the wire. The front of this wave travels with a velocity corresponding to the gradient of BC in Fig. 39, i.e. with the velocity of elastic waves in the material. It consequently eventually overtakes the plastic wave front. In Fig. 40 (b) the stress, strain, and particle velocity distributions soon after unloading are shown. Between the end of the wire and the front of the wave of unloading the stress has disappeared, the strain has been reduced by an amount corresponding to the elastic deformation (CD in Fig. 39), and the particle velocity has fallen by an amount $\sigma_1/(\rho c_0)$. In Fig. 40 (c) the front of the wave of unloading has just reached the front of the plastic wave and the particle velocity is $V_1 - \sigma_1/\rho c_0$ in the unstressed region up to F and $\sigma_0/\rho c_0$ between F and the front of the elastic wave.

At this stage a plastic wave of reduced amplitude will move

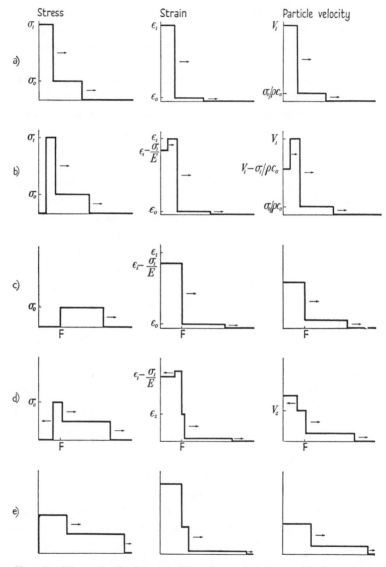

Fig. 40. Wave of unloading travelling along wire of material whose stress-strain curve is that shown in Fig. 39. In (d), after internal reflection at F,
$$\sigma_2 = \rho c_0 (V_1 - V_2) - \sigma_1 \text{ and } \epsilon_2 = \sigma_0/E + (\sigma_2 - \sigma_0)/S_1.$$

forward along the wire from F and an elastic wave will move back, the effect being in the nature of an 'internal reflection' at F. The two waves generated at this reflection will both be waves of tension and the particle velocities on both sides of F will be equal. From the condition that the values of the stress and of the velocity on either side of F are equal after reflection, the amplitudes of the two waves generated may be deduced. Fig. 40 (d) shows a plastic wave of reduced amplitude moving along the wire from F whilst the reflected elastic wave travels back towards the end of the wire. In 40 (e) it has reached the end of the wire and conditions of stress and velocity are similar to 40 (a), except that the particle velocity between the end of the wire and the plastic wave front has lower values. The whole cycle is then repeated, and when the second compression wave travels down the wire and reaches the front of the plastic wave it reduces its amplitude once again, so that the residual strain in the wire has a step-like character, each step corresponding to a point at which an elastic wave of compression has reached the head of the plastic wave.

For Fig. 40 the gradient of the stress-strain curve in the plastic region has been taken as one-ninth of its value in the elastic region, so that $E = 9S_1$ and $C = 3c_0$. Further, σ_1 has been taken as $4\sigma_0$ so that $\epsilon_1 = 28\epsilon_0$ (see Fig. 39). In deriving these curves it is simplest to start with the one of particle velocity, since once the end of the wire is released the momentum will remain constant and this means that the area under the particle velocity curve must also be constant. This provides a useful check of the conditions after the internal reflection at F.

Plastic waves in specimens of finite length

The plastic waves so far considered in this chapter are those produced when a wire is stretched beyond its elastic limit. Exactly the same analysis applies to the problem of sudden compression, and the theory has been applied to the impact of bars by White (1949) and DeJuhasz (1949). If one end of a bar is suddenly compressed beyond its elastic limit and the stress is maintained an elastic wave of compression will travel along the bar, and this will be followed by a plastic wave which travels more slowly. When the stress is removed a wave of unloading, which in this case is a wave of tension, travels along the bar at a velocity greater than that of the plastic wave and when it reaches the plastic wave front it reduces the amplitude of the plastic wave as shown in the previous section. With a bar of finite length the elastic wave is reflected at the other end, and if this end is fixed a plastic wave is generated on reflection. Thus if one end of a bar is compressed for a short time and then released a number of different waves travel in it in both directions, and the stress distribution some time after impact becomes extremely complicated.

White (1949) has considered the problem for a material which has a stress-strain relation of the type shown in Fig. 39, and has illustrated the propagation of the fronts of the various waves on an (x, t) diagram. Such a diagram is shown in Fig. 41, for a bar which is struck at one end, the other end being fixed. Elastic waves are shown in the figure as thin lines and plastic waves as thick lines. The length of the bar is assumed to be OL and a constant compressive stress is applied for a time OT and then removed. The (x, t) relation for the front of the original elastic wave is denoted by OA and the relation for the front of the plastic wave by OP. From T the wave of unloading travels with the velocity of elastic waves and meets the plastic wave at P_1; an elastic wave of compression then travels back to the end of the bar whilst the plastic wave travels on with reduced amplitude but with the same velocity of propagation to P_2, where it once again meets an elastic wave which has been reflected at the end of the bar, and the same process is repeated at P_3, P_4, etc., the amplitude of the plastic wave being reduced on each occasion. In the meantime the elastic wave has reached the fixed end of the bar at A, and since, at the instant of reflection, the stress is at the proportional limit everywhere between the head of the plastic wave and the fixed end of the bar, the excess stress built up on reflection is transmitted back as a plastic wave, and is shown in the figure travelling back along the bar with the velocity of plastic waves along AB. It meets the oncoming plastic wave at C and there will be an interaction between the two plastic wave fronts at this point, resulting in a further increase in the strain.

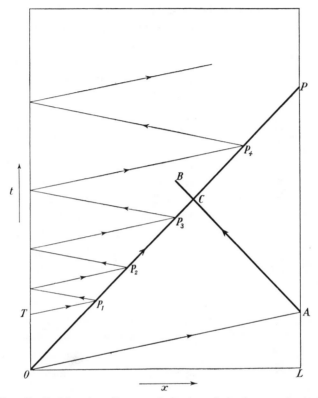

FIG. 41. Position-time diagram of plastic and elastic wave fronts in bar of finite length. Thin lines denote elastic wave fronts and thick lines plastic wave fronts.

Experiments on plastic waves

There are so far very few published data of experiments on the propagation of plastic waves. von Kármán and Duwez (1950) describe some experiments on the propagation along copper wires, whilst White (1949) discusses some experiments on the impact of steel bars carried out at the California Institute of Technology. The apparatus used by Duwez for testing the theory of the propagation of plastic waves is shown in Fig. 42.

Annealed copper wires were used and equidistant marks were made along them before deformation. By measuring the separation between these marks after the experiment the distribution of strain could be deduced. The end of the wire is attached

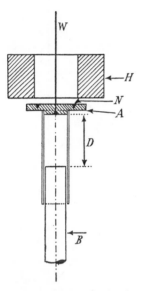

to the rigid member A. B is a vertical rod which rests on the bottom frame of the testing machine, its upper end fitting loosely into the tubular part of A. H is a hammer which falls between two vertical rails, and is accelerated by pre-stretched rubber bands; when it hits A the wire is elongated until the disk on A reaches the rod B. The disk on A has a circular notch at N and the rim breaks off at this notch when A has travelled the distance D, and is stopped by B. The hammer then continues downwards. The time of impact can be varied by changing the distance D, and the velocity of the hammer can also be varied. The hammer is sufficiently massive not to be decelerated appreciably while the wire is being stretched, and its velocity is measured electrically.

Fig. 42. Apparatus for investigating the propagation of plastic waves in a wire.

The purpose of the experiments was to verify three points about the propagation of plastic waves, and separate experiments were carried out with this object. The three points were:

(1) the existence of a plastic wave front of given magnitude;
(2) the relation between the velocity of impact V_1 and the strain ϵ_1 behind the plastic wave front;
(3) the distribution of plastic strain between the plastic wave and the elastic wave.

In order to test (1) the velocity of the hammer was kept constant and the duration of the impact was varied by changing the distance D. A region of constant strain, the length of which was proportional to the duration of the impact, was found at the end of the wire in these experiments. As predicted by the theory, a plastic wave the magnitude of which is constant at constant velocity is thus propagated down the wire.

In order to test relation (2) the velocity of the hammer (V_1) was varied and the resulting plastic strain (ϵ_1) at the end of the wire was measured. The theoretical relation between V_1 and ϵ_1 is given by equation (7.11), and this was calculated for the wire by measuring the stress-strain curve statically. The (S, ϵ) curve could then be observed and integrated numerically. The agreement between theory and experiment was found to be quite good.

In testing the third point, viz. the distribution of plastic strain between the plastic wave front and the elastic wave front, Duwez assumed that at the end of the impact the motion ceased and the plastic deformation was 'frozen' in the wire. As shown earlier in this chapter this is not in fact what occurs, since at the end of the impact a wave of unloading travels down the wire. When it reaches the plastic wave front it reduces the amplitude of the latter and is itself reflected back to the end of the wire. A series of such reflections back and forth along the wire occur so that the final distribution of plastic strain beyond the plastic wave front is far more gradual than it was at the instant at which the impact ceased. Further, the curve of strain against distance might be expected to be step-like in shape, each step corresponding to a point where an elastic wave of unloading reaches the front of the plastic wave. Duwez found experimentally that the distribution of plastic strain in front of the plastic wave was in fact flatter than he had expected, and subsequent calculations in terms of the multiple reflection of waves of unloading have given better agreement with the experimental results.

The other experimental work on plastic waves described by White (1949) was carried out by Duwez and Clark on steel specimens both in tension and compression. The interpretation of these results is complicated by the fact that the proportional limit of this material is highly sensitive to the rate of loading, the yield stress under dynamic conditions being two or three times as great as it is statically. Until the dynamic stress-strain relations of such materials have been investigated by other means experiments on the propagation of plastic waves in them is unlikely to prove very profitable.

Shock waves in solids

As shown in an earlier section of this chapter on the treatment of plastic waves by the Eulerian method, the equations of motion and continuity in a solid rod or wire are formally similar to the equation of a wave of finite amplitude in a fluid. The velocity of propagation of a disturbance is given by equation (7.21) as $c + V$ and, if the elastic modulus $S = d\sigma'/d\epsilon$ is constant, large *compressive* disturbances will travel faster than smaller ones so that any finite compression pulse will eventually acquire a steep front as it travels through the medium. In solids the particle velocity, even for intense disturbances, is very small compared with the velocity of propagation so that, if S is constant, stress pulses can travel for considerable distances without change in form, and it is changes in the value of this elastic modulus S with strain that are principally responsible for the distortion of pulses of finite amplitude. For most solids S decreases beyond the proportional limit, and plastic waves rather than shock waves are set up in rods of the material when the deformations are sufficiently large. There are some solids, however, such as rubbers and other high polymers, where large tensile strains result in an orientation of the long chain molecules and this is accompanied by a large increase in the value of $d\sigma'/d\epsilon$. It would therefore appear that when large deformations are propagated through these materials shock waves may develop. Up to the present time no experimental work appears to have been carried out to investigate whether this does in fact occur.

Another condition in which shock waves may be set up in solids is when dilatational waves of large amplitude are propagated through them. As shown in Part I of this monograph, elastic waves of dilatation are propagated through a solid with the velocity $[(\lambda+2\mu)/\rho]^{\frac{1}{2}}$, or when expressed in terms of the bulk modulus k the velocity is $[(k+\frac{4}{3}\mu)/\rho]^{\frac{1}{2}}$. Now as shown by Bridgman (1931), the compressibility k of solids increases at high pressures, and as a result of this the velocity of bulk waves of large amplitude might be expected to be greater than that of waves of small amplitude. The shear modulus μ is unlikely to be of importance since long before such high pressures are attained the shear yield point will have been passed and the material will in fact be behaving like a fluid.

Bridgman (1931) has shown that the relation between the volume v' and the hydrostatic pressure P in a solid can be represented by the relation

$$\frac{v'_0-v'}{v'_0} = aP-bP^2, \qquad (7.22)$$

where v'_0 is the original volume and a and b are constants. If P is measured in kilograms per square centimetre a is of the order of 5×10^{-7} and b is of the order of 10^{-12} for most metals. (For pure iron $a = 5\cdot9\times10^{-7}$ and $b = 2\cdot1\times10^{-12}$.)

The bulk modulus k is $v'\,\partial P/\partial v'$, and at low pressures it approaches the constant value $1/a$, whilst at higher pressures it is given by $1/(a+2bP)$. It may thus be seen that for metals a pressure of the order of 1,000 kg./sq. cm. will be required to change the compressibility by 1 per cent. and this will result in a change of only $\frac{1}{2}$ per cent. in the velocity of propagation. Clearly the curvature in the (P,v) diagram of metals will only become important at very high pressures, and shock waves are thus only likely to be produced in a solid either when it is in direct contact with a detonating explosive or when a high-speed projectile is fired at it.

It is beyond the scope of this monograph to deal in any detail with the theory of shock waves, but full discussions of the subject will be found in Taylor and Maccoll (1935), Herpin (1947), Courant and Friedrichs (1948), and Penney and Pike (1950). We shall, however, give a very brief account of how the fundamental shock wave equations are obtained. These relations, which are known as the Rankine–Hugoniot equations, are derived from the equations for the conservation of mass, momentum, and energy in the medium.

In deriving these equations we assume that a plane shock wave has been set up and is travelling through the material with a constant velocity c. In the region behind the shock front the particle velocity, pressure, and density are assumed constant; there is a transition zone at the front of the shock wave, and

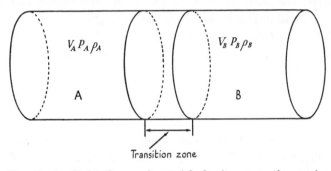

Fig. 43. A cylindrical mass of material of unit cross-section passing from right to left through a shock transition zone. System of reference chosen so that transition zone appears at rest.

then undisturbed material. The conditions in the transition zone are also assumed to be steady. We now choose our system of reference so that the transition zone appears to be at rest. Fig. 43 shows a cylinder of material of unit cross-section which contains the transition zone. In the system of reference chosen let the pressure, density, and particle velocity behind the transition zone in the region A be P_A, ρ_A, and V_A, and in the undisturbed material B let these quantities be P_B, ρ_B, and V_B.

Since no matter is accumulating anywhere, the mass of material entering the transition zone per unit time must be equal to the mass of material leaving it, and if we call this mass m we have

$$m = \rho_A V_A = \rho_B V_B. \tag{7.23}$$

The material enters the transition zone with velocity V_B and leaves with velocity V_A; equating the rate of change of momentum to the force acting we obtain

$$m(V_B - V_A) = P_A - P_B. \tag{7.24}$$

Finally the rate at which work is being done by the cylinder is $P_A V_A - P_B V_B$ per unit time, and this is equal to the change in the kinetic energy and in the internal energy of the material in passing through the transition zone. If we call the change in the internal energy per unit mass ΔU we have

$$P_A V_A - P_B V_B = m[\tfrac{1}{2}(V_B^2 - V_A^2) + \Delta U]. \tag{7.25}$$

(7.23), (7.24), and (7.25) are the three equations from which the Rankine–Hugoniot relations may be derived.

Thus from (7.23) and (7.24) we find that

$$V_B = v_B[(P_A - P_B)/(v_B - v_A)]^{\frac{1}{2}}, \tag{7.26}$$

where $v_A = 1/\rho_A$ and $v_B = 1/\rho_B$. Now V_B is the velocity which the material in the undisturbed region appears to have if the system of reference is chosen to make the transition zone stationary. It is therefore equal to the velocity of propagation of the shock front c through undisturbed material.

The particle velocity behind the shock front relative to the undisturbed material is $V_B - V_A$, and from (7.23) and (7.24) this is given by

$$V = V_B - V_A = [(P_A - P_B)(v_B - v_A)]^{\frac{1}{2}}. \tag{7.27}$$

Finally we obtain from (7.25) the relation

$$\Delta U = \tfrac{1}{2}(P_A + P_B)(v_B - v_A). \tag{7.28}$$

The three equations (7.26), (7.27), and (7.28) are the Rankine–Hugoniot relations for a travelling shock front, and the first two of these equations are derived purely from the conservation of mass and momentum and therefore apply even when chemical energy is being generated in the medium, as in a detonation wave passing through an explosive. From (7.26) it may be seen that for very small pressure differences the velocity c approaches the velocity of sound in the medium, whilst from (7.26) and (7.27) the relation between pressure difference ΔP and the particle velocity V becomes

$$\Delta P = \rho V c, \tag{7.29}$$

which is the relation obtained for plane elastic waves (cf. equation (3.8)).

The thickness of the transition zone depends on the properties of the medium and is governed by the dissipative forces which become important as the velocity gradients in this zone increase. In the case of fluids it has been shown to be of the order of one mean free path of the molecules. For gases this is clearly a limiting thickness, since there must be a separation of at least this order between the molecules which have been accelerated by the oncoming shock front and those which have not yet been affected. No work appears to have been published on the probable thickness of the transition zone in solids.

CHAPTER VIII

FRACTURES PRODUCED BY STRESS WAVES

WHEN a stress pulse of sufficiently large amplitude travels through a solid it may produce fractures, and the purpose of this chapter is to describe the manner in which such fractures are formed and the way in which they differ from those produced under conditions of static loading. The fractures produced by stress pulses differ from those produced 'statically' for several different reasons. First, since the velocity of crack propagation is generally considerably lower than the velocity of propagation of the pulse (see Edgerton and Barstow (1939–41), Schardin (1950), Christie (1952)) for pulses of short duration any cracks that are formed do not have time to grow before the pulse has passed on and the stress has been removed. Secondly, with a short pulse only a small part of the specimen is stressed at any one time and fractures may form in one region of a specimen quite independently of what may be occurring elsewhere. Thirdly, as shown in Chapter II, when a compression pulse is incident on a free boundary it gives rise to a reflected tension pulse, whilst when it is reflected obliquely both a dilatational and a distortional pulse are produced. The interference of such reflected pulses may, as shown in Plate I, give rise to very complicated stress distributions and the superposition of several reflected pulses may produce stresses which are sufficiently large to cause fracture when the amplitude of the incident pulse was too small to do so. Lastly, as shown in Chapter IV, the dynamic elastic behaviour of many solids may differ considerably from that observed statically, and at the very high rates of loading associated with intense stress pulses, materials which are normally regarded as ductile may behave in a brittle manner.

J. Hopkinson's experiments

Among the earliest experimental work on the fractures produced by stress waves was that carried out by J. Hopkinson (1872) who measured the strength of steel wires when they were suddenly stretched by a falling weight. The apparatus he used was in principle similar to that described recently by von Kármán and Duwez (1950) and shown in Fig. 42. A ball-shaped weight pierced by a hole was threaded on the wire and was dropped down the wire from a known height so that it struck a clamp attached to the bottom of the wire. Hopkinson used different weights dropped from various heights and obtained the rather remarkable result that the minimum height from which a weight had to be dropped in order to break the wire was independent of the size of the weight. In other words, under these conditions the effects of two blows were equivalent not when their momenta or energies were the same but when their velocities were equal. J. Hopkinson explained this result in terms of the propagation of elastic waves up and down the wire. When the weight hits the clamp the end of the wire acquires a particle velocity V_0 equal to that of the weight and clamp, and a steep-fronted wave of tension is propagated up the wire with the velocity of extensional elastic waves, $c_0 = (E/\rho)^{\frac{1}{2}}$. The stress σ_0 at the head of the pulse is given by the equation

$$\sigma_0 = \rho V_0 c_0 \quad \text{(see equation (3.8), Chapter III).} \quad (8.1)$$

This stress acts on the weight and decelerates it; the equation of motion for the weight if it is of mass M is

$$M \frac{dV}{dt} = A\sigma = \rho A V c_0. \quad (8.2)$$

V is the velocity of the weight at time t, σ is the stress at the bottom of the wire, and A is the cross-sectional area of the wire. The solution of (8.2) is

$$V = V_0 \exp \frac{-\rho A c_0 t}{M}. \quad (8.3)$$

The stress at the bottom of the wire at time t can be seen from (8.2) and (8.3) to be given by

$$\sigma = \rho c_0 V_0 \exp \frac{-\rho A c_0 t}{M}, \quad (8.4)$$

and thus the wave travelling up the wire has a sharp front of magnitude $\rho V_0 c_0$, and the tension decreases exponentially with distance behind the wave front, the stress at distance x from the weight being given by

$$\sigma = \rho c_0 V_0 \exp\left[\frac{\rho A}{M}(x - c_0 t)\right]. \qquad (8.5)$$

When this wave reaches the top of the wire it is reflected as a wave of tension so that immediately after reflection has commenced the tensile stress at the top of the wire has twice the value it had at the head of the oncoming pulse. The reflected part of the pulse travels back through the tail of the incident pulse and is reflected again at the weight. The pulse thus continues to travel up and down the wire. In J. Hopkinson's experiments the head of the stress wave was able to traverse the length of the wire several times before the weight M had been decelerated sufficiently for the stress to have fallen to a fraction of its value at the head of the pulse. Consequently different parts of the wave were travelling in opposite directions along the wire simultaneously, and the resultant stress distribution was very complex. If the stress at the first reflection ($= 2\rho V_0 c_0$) is sufficient to break the wire, fracture would be expected to take place very near the top end of the wire, and J. Hopkinson found in his experiments that this in fact was where the wire generally broke. Further, under these conditions fracture should occur however small M might be, the stress depending only on the value of V_0.

B. Hopkinson (1905) repeated his father's experiments with an apparatus which enabled him to measure the maximum strain at the top of the wire, also he used smaller weights so that the rate of exponential decay in the tail of the stress wave was very much more rapid. Nevertheless, as shown by Taylor (1946), the maximum tensile stress in B. Hopkinson's experiments did not occur at the first reflection when the stress was $2\rho V_0 c_0$, but at the third reflection, i.e. the second reflection at the top of the wire, when the stress reached $2 \cdot 15 \rho V_0 c_0$. B. Hopkinson in these experiments showed that the tensile strength of metal wires was very much greater under these dynamic conditions than when measured statically, and the correction for the stress made by Taylor only reinforces this conclusion.

Fractures produced by explosives

B. Hopkinson (1912) investigated the fractures produced in metal specimens when small quantities of explosive were detonated in contact with them. He used steel specimens and the charges of gun-cotton were detonated in contact with plates of various thickness. With thin plates (less than $\frac{1}{2}''$ in thickness) Hopkinson found that the explosive punched a hole right through the plate. With thicker plates, however, the explosive was found to dent the surface of the plate with which it was in contact and to throw off a circular disk of metal from the opposite face. Hopkinson termed this disk of metal a 'scab', and the surface of this scab which had broken away from the metal plate was of irregular texture and roughly in the shape of a spherical cap.

When an explosive is detonated it is converted to hot gas in a few microseconds, the pressure rising very rapidly to many tens of tons per square inch. Since in these experiments the hot explosive gases are confined only by the surrounding air they expand very rapidly and the pressure falls off in a few hundred thousandths of a second. Consequently a spherical compression pulse of very short duration is sent into the steel specimen. This spherical pulse arrives at the bottom face of the specimen and is reflected as a pulse of tension and since the steel is very much weaker under tension than under compression it is this reflected tension pulse which produces the scab in the material.

The effect is shown diagrammatically in Fig. 44. The charge is situated at P on the top surface of the plate AB and the spherical wave of compression travels out from it and is reflected at the free bottom face of the plate CD. In the region immediately beneath the charge it is reflected entirely as a wave of tension; farther out it will approach the bottom face at oblique incidence and a reflected distortional wave will be generated in addition to the wave of tension. (This effect was discussed in Chapter II and Fig. 6 shows the related amplitudes of these waves for different angles of incidence for a material for which Poisson's ratio is 0·33.)

A dilatational wave is always reflected at an angle equal to the angle of incidence so that it appears to be diverging from the point P' which is the image of the point P in the plane CD. Now this reflected tension pulse will pass through the tail of the incident compression pulse so that the resultant stress in the plate will be the sum of the stresses due to the incident and reflected pulses. The stress distribution when a compression

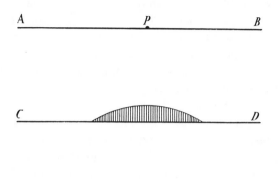

FIG. 44. Formation of a 'scab' in a large sheet of material.

pulse is incident normally on a free surface is discussed in Chapter IV and is shown diagrammatically in Fig. 21. The shape of the pressure pulse obtained with gun-cotton will not necessarily be that shown in Fig. 21, although as a result of the very rapid detonation and relatively slower escape of the detonation products the pulse might be expected to have a sharp rise and a more gradual fall. Further, as a result of the large pressures set up on detonation the wave will probably develop a shock front which will persist in its passage through the specimen at least in the region near the charge.

As shown in Fig. 21, appreciable tension will first be set up some distance from the free surface and it is here that fractures will begin. Once a fracture has started the rest of the pulse is reflected at the new free surface formed, so that a series of parallel cracks may be produced when the amplitude of the stress pulse is sufficiently great. Each time a fracture is formed a certain amount of forward momentum is trapped between the fracture and the bottom surface, and if the fractures extend sufficiently for a piece of the metal to be broken off this will fly off with the momentum trapped in it. This phenomenon is similar to that used in the Hopkinson bar (q.v.) where the shape of the pressure pulse is measured by trapping sections of it in a *time-piece* and measuring the momentum that the time-piece has acquired. In Hopkinson's experiments on scabbing, described above, he found that the scab flew off with considerable velocity, and was in fact able to penetrate a thick wooden plank.

The region in which fracture occurs, which is shown shaded in Fig. 44, is limited to those parts of the specimen where the tensile stress of the reflected dilatation wave is greater than the tensile strength of the material. The amplitude of this reflected tension pulse may decrease as a result of four factors:

1. Since the pulse is spherical its amplitude decreases with the radius, which in Fig. 44 is the distance from the image point P'.

2. On moving outwards from the line PP' the angle of incidence of the incident compression pulse increases. It was shown in Chapter II that as the angle of incidence is increased the amplitude of the reflected dilatation wave decreases, the remaining energy going into the reflected distortional wave. (It may be seen from Fig. 6 that as oblique incidence is approached the amplitude of the reflected dilatational wave rises once again; for $\nu = 0.33$ the angle of incidence for minimum amplitude is about 70°.)

3. As a result of internal friction some of the elastic energy of the pulse is converted into heat as it passes through the specimen and this results in a lengthening of the pulse and a decrease in its amplitude, the momentum of the pulse being conserved.

4. When fracture occurs the amplitude decreases rapidly as the elastic energy stored in the material around the fracture is dissipated.

As a result of the above considerations the fracture region might be expected to be very roughly in the form of a spherical cap with P' as centre, and in experiments with more localized charges than those used by Hopkinson which are described in the next section the scab was found to be approximately of this shape.

In a recent paper Rinehart (1951) has repeated Hopkinson's experiments using cylindrical charges of explosive in contact with steel plates. After detonation the sheets were sectioned and examined for changes in the hardness and the micro-structure of the metal. Rinehart has shown that in addition to the fractures found by Hopkinson the stress pulse causes a con-siderable amount of plastic distortion in the metal and changes its micro-structure, these effects being produced by the shear stresses set up by the pulse.

Experiments with specimens of different shapes

In the experiments described above only the stress pulse reflected from the bottom face of the plate was of sufficient amplitude to produce fractures, the sides of the plate being too far away from the charge for the waves reflected from them to fracture the specimen. With smaller specimens, however, the waves reflected from the sides produce fractures and there is also reinforcement between the waves from the sides and the bottom face of the specimen which may result in additional fractures. Shearman, Christie, and the author (Kolsky and Shearman, 1949; Kolsky and Christie, 1952) have carried out experiments with specimens of transparent materials in order to observe the fractures produced by the interference between reflected pressure pulses.

Specimens of plastics and glass were used and a small charge of lead azide (up to 0·5 g.) was painted on to the surface of the specimen in the form of a small, roughly hemispherical mound. A speck of silver fulminate was placed on top of the charge to ensure rapid detonation and the charges were fired with an electrically heated wire. Most of the experiments were done with specimens of 'Perspex' (a plasticized polymethyl-methacrylate) since this material can be obtained fairly free from non-uniform internal strains and the velocity of propagation of dilatational waves in it is comparatively low (about 2,000 metres per second) so that the length of a pulse of 2 microseconds duration is only 4 mm. Since the effective duration of the pulses obtained with the very small charges employed was of this order, the dimensions of the specimens could be made large compared with the pulse length. This simplified the nature of the stress distribution obtained after reflection.

When a charge is detonated on the surface of a large block of material a small surface crater is formed. The crater is surrounded by a small, roughly hemispherical region of fracture. This is produced by the tensile stresses which are set up by the outgoing spherical compression pulse parallel to the wave front. The damage consists of a very large number of minute hair-like cracks radiating from the seat of the charge. (This effect can be seen on Plate III in the photograph of the fractures formed in a conical specimen. The hemispherical region is here around the centre of the base of the cone.) If the specimen is sufficiently large this is the only damage observed, since the amplitude of the outgoing spherical stress pulse is too small when it reaches the free surface of the specimen to produce fractures.

When a charge is detonated on the surface of a sheet of material remote from the edges, a scab is formed on the opposite face in the region immediately beneath the charge. This effect was discussed in the previous section in describing B. Hopkinson's experiments on steel specimens. If the charge is near the edge of the sheet reflected tension pulses will be produced both from the bottom surface and the side. The pulse reflected from the bottom face causes a scab and the pulse reflected from the side causes a surface fracture parallel to the edge. In the experiments with small lead azide charges on specimens of plastics this latter fracture was found to occur a few millimetres from the edge. The distance of this fracture from the edge will depend on the amplitude and shape of the outgoing pressure pulse. As shown in Fig. 21 this distance should never exceed half the pulse length, and for a steep-fronted pulse it should become equal to half the pulse length when the amplitude is just sufficient to fracture the material.

In addition to the fractures caused by the separate reflections of the pulse from the bottom face and side of the specimen a fracture may occur as a result of the reinforcement of the two reflected pulses. The effect is shown diagrammatically in Fig. 45, which represents the cross-section of a rectangular specimen. The charge is at P on the top surface. BC and AD are sections of the side surfaces, CD being in the bottom surface. The

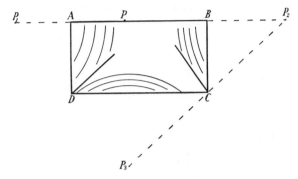

FIG. 45. Reinforcement between pulses reflected from the sides and bottom face of a small rectangular specimen.

dilatational waves reflected from BC and CD appear to originate from the image points P_2 and P_3, and reinforcement between them will therefore occur along the plane containing C which is normal to $P_2 P_3$, the two reflected tension pulses arriving simultaneously at any point in this plane. A similar reinforcement will occur in the plane containing D which is perpendicular to $P_1 P_3$. In experiments on small rectangular specimens fracture surfaces were found to occur along these planes of reinforcement and Plate II shows the top and side views of a square specimen in which these fractures may be seen.

When a charge is detonated at the centre of one of the end faces of a cylindrical specimen a number of different fracture regions are formed and these are shown diagrammatically in Fig. 46. The different fractures are:

(a) A small region of compression damage around the seat of the charge P which is shown by the shaded area in Fig. 46.

(b) A circular crack on the top surface a few millimetres from the edge. This is shown by S and T in the figure and results from the reflection of the compression pulse at the cylindrical surface of the specimen as a wave of tension.

(c) A linear region of fracture extending for some distance down *PC*, the axis of the cylinder. This fracture region is produced by the wave reflected from the curved surface converging on to the axis of the cylinder so that a very large tension is built up in this region. The tensile stress is applied radially and if the charge were actually a point source the tension at the axis would become infinite. Since the charge is of finite size a large tensile stress is applied in a region round the axis, the radius of which is comparable with that of the charge. The length of this fracture region *PC* was found to increase with the weight of explosive used, the end *C* corresponding to the point where the maximum resultant tension was just equal to the tensile strength of the plastic.

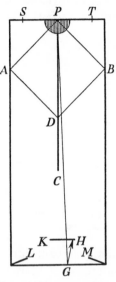

Fig. 46. Diagram of fracture regions in a cylindrical specimen when a charge is detonated at the centre of one end face.

The intensity of the fracture along *PC* was found not to fall off uniformly but to pass through a maximum and then decrease rapidly. The reason for this would appear to be that for a 'ray' *PA* which is reflected along *AD* the intensity at *D* will depend both on the length *PA+AD* through which the wave has travelled and on the angle of incidence of *PA* on the cylindrical surface. For points farther down the axis of the cylinder the distance travelled and the angle of incidence both increase, but whilst the amplitude is inversely proportional to the distance, the effect of increasing the angle of incidence is first to decrease and then to increase the amplitude of the reflected wave of tension (see Fig. 6).

(When a charge is detonated at the centre of the base of a conical specimen fractures are formed along the axis of the cone

PLATE II

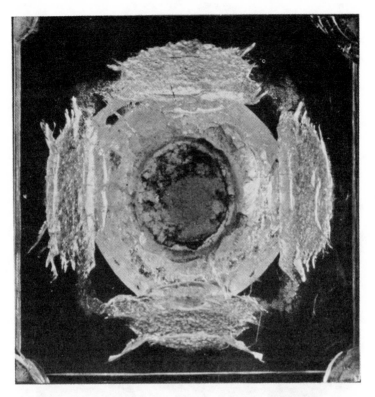

(a)

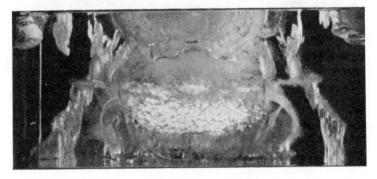

(b)

Fractures produced in a square 'Perspex' specimen by 0·3 g. of lead azide.
(a) Top view. (b) Side view.

(Magnification × 3)

PLATE III

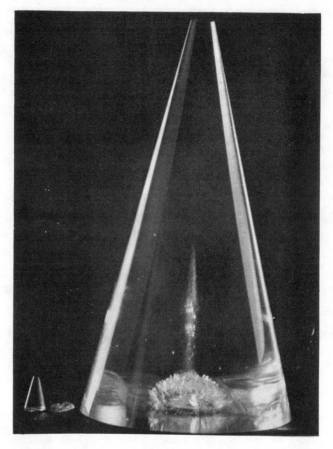

Fractures in a 'Perspex' cone produced by 0·4 g. of lead azide.
(Actual size)

in the same way as in cylinders, and the photograph of the conical specimen in Plate III shows this region of fracture.)

(d) With short cylinders a flat region of fracture HK, parallel to the base of the cylinder, was found to occur. This results from the reflection of the pulse at the flat end of the cylinder and is similar to scab formation in flat sheets. The ray PGH is shown in the figure to illustrate the formation of this fracture.

(e) Finally the reinforcement of the pulse reflected at the base of the cylinder by that reflected from the bottom of the cylindrical surface results in a conical fracture surface shown by L and M in the figure. This effect has already been discussed and is illustrated by Fig. 43. It may be seen that the tangent of the angle that this conical region of fracture makes with the base of the cylinder will be equal to the ratio of the radius of the cylinder to its height.

Fractures in conical specimens

The fractures described above can all be accounted for in terms of the reflection of the outgoing pressure pulse at the free surfaces of the specimen, the propagation being analogous to that described by geometrical optics in the study of light. When a pulse is propagated along the axis of a conical specimen, however, it increases in amplitude and changes in shape as it progresses towards the apex. The theory of the propagation of a longitudinal elastic pulse along a cone when the cross-section is small compared with the length of the pulse was considered in Chapter III. It was shown there that if the pulse is originally in the form of compression it develops a tension tail, the length of the compression region becoming shorter and shorter as the apex is approached. At the same time the maximum stress amplitude in the tension tail increases, so that if the material is very much stronger in compression than it is in tension the tip may break in tension before the head of the pulse reaches the apex of the cone.

When a lead azide charge is detonated at the centre of the base of a cone of plastic the tip is in fact found to fly off with high velocity and the result of such an experiment is shown in Plate III. In addition to the tip a small disk-shaped fragment which fitted between the tip and the rest of the cone was recovered, and this is shown in the photograph. The momentum with which the tip leaves the cone corresponds to the section of the pressure pulse trapped in it, and the phenomenon is similar to that which occurs in the Hopkinson pressure bar (see Chapter IV). An effect similar to the fractures at the tips of cones was found to occur in specimens which were in the form of small sheets with sharp corners. The outgoing pressure pulse will here converge on the corner, the problem being a two-dimensional analogue of the propagation in a cone. Such fractures can be seen at the four corners of the square specimen shown in Plate II.

Type of fracture surface produced by stress pulses

We have so far considered the way in which the geometrical shape of a specimen affects the positions of fractures formed in it by pressure pulses. As was mentioned at the beginning of this chapter, the fracture surfaces themselves are different under these conditions from those obtained when the specimen is stressed statically. This is because the stress is applied for so short a time that any cracks formed do not have time to spread, and instead of running cracks, a large number of separate fractures occur, and these sometimes join up to form a more or less continuous surface which is irregular in texture. Such surfaces of fracture may be seen to have formed inside the square specimen shown in Plate II. Also, the fractures along the axis of the conical specimen may be seen to consist of a series of 'bubbles' rather than of running cracks, their appearance being more analogous to cavitation in a liquid than fracture in a solid.

The pulses produced by the small lead azide charges used are of the order of 2–3 microseconds in duration, and since the velocity of crack propagation in glass and plastics is less than 1,500 metres per second (see Edgerton and Barstow, 1939–41, and Christie, 1952) a crack would not be able to spread more than a few millimetres before the pulse has passed on and the stress has been removed. When an explosive which produces a pulse of longer duration is employed the fractures become more like those observed by static loading. Thus when a small nitroglycerine charge was detonated on the surface of a sheet of 'Perspex' the scab formed was very irregular in shape and had a conchoidal appearance. The reason for this irregular shape is that the length of the pulse in the plastic was several times the thickness of the plate, so that different sections of the pulse were travelling back and forth across the plate simultaneously and the resultant stress distribution was similar to that in J. Hopkinson's experiments with steel wires. Further, because in this experiment the duration of the pulse was very much longer, cracks were able to grow to a length of several centimetres before the stress was removed; consequently smooth fracture surfaces characteristic of growing cracks were formed in the specimen.

Brittle and ductile fracture

As mentioned earlier the fractures produced by sharp stress pulses may also differ from those produced statically because of changes in the mechanical behaviour of solids at high rates of loading. These differences are not associated with the propagation of stress waves as such, and occur whenever the rate of loading is sufficiently high. With ductile solids the effect of increasing the rate of loading is that the fractures formed become more like those found in brittle materials. This problem has been discussed by B. Hopkinson (1910) and more recently by Lethersich (1948). Ductility is associated with the flow of the solid under an applied shear stress whilst brittle fracture occurs when minute cracks grow under an applied tensile stress. When a force is applied for only a very short time the shear stresses built up do not have time to produce an appreciable amount of flow and many materials consequently withstand momentary stresses of much greater magnitude than their static yield stress (see Taylor, 1946). Further, under these conditions when failure does occur it is in the form of brittle fractures with no flow around the fractured surfaces. In the experiments with 'Perspex' specimens described earlier in the chapter this effect was investigated by viewing the fractured specimens in polarized light. When this plastic is deformed slowly a large amount of residual strain is found to persist after the removal of the load. The specimens on which small explosive charges had been detonated did not, however, show such residual strain even in regions immediately surrounding the fractured surfaces.

APPENDIX

Notations for stress, strain, and elastic constants

MANY different systems have been employed for denoting the components of stress and strain and the notation in the text is compared below with the two notations which are most commonly used in the literature, namely that used by Love (1928) and that used by von Kármán (1910). The former has also been used by Southwell (1941) and Planck (1932), whilst the latter is also used by Timoshenko (1934).

	Text	von Kármán	Love
Cartesian coordinates			
Normal stresses	σ_{xx}, σ_{yy}, σ_{zz}	σ_x, σ_y, σ_z	X_x, Y_y, Z_z
Shearing stresses	σ_{yz}, σ_{zx}, σ_{xy}	τ_{yz}, τ_{zx}, τ_{xy}	Y_z, Z_x, X_y
Normal strains	ϵ_{xx}, ϵ_{yy}, ϵ_{zz}	ϵ_x, ϵ_y, ϵ_z	e_{xx}, e_{yy}, e_{zz}
Shear strains	ϵ_{yz}, ϵ_{zx}, ϵ_{xy}	γ_{yz}, γ_{zx}, γ_{xy}	e_{yz}, e_{zx}, e_{xy}
Rotation	$\bar{\omega}_x$, $\bar{\omega}_y$, $\bar{\omega}_z$	..	$\bar{\omega}_x$, $\bar{\omega}_y$, $\bar{\omega}_z$
Dilatation	Δ	e	Δ
Cylindrical coordinates			
Normal stresses	σ_{rr}, $\sigma_{\theta\theta}$, σ_{zz}	σ_r, σ_θ, σ_z	$\widehat{rr}$, $\widehat{\theta\theta}$, $\widehat{zz}$
Shearing stresses	$\sigma_{\theta z}$, σ_{zr}, $\sigma_{r\theta}$	$\tau_{\theta z}$, τ_{zr}, $\tau_{r\theta}$	$\widehat{\theta z}$, $\widehat{zr}$, $\widehat{r\theta}$
Normal strains	ϵ_{rr}, $\epsilon_{\theta\theta}$, ϵ_{zz}	ϵ_r, ϵ_θ, ϵ_z	e_{rr}, $e_{\theta\theta}$, e_{zz}
Elastic constants			
Young's modulus	E	E	E
Bulk modulus	k	k	k
Poisson's ratio	ν	ν	σ
Rigidity modulus	μ	μ or G	μ
Lamé's constant	λ	λ	λ

The tensor notation for stress and strain provides a very convenient method of expressing elastic relations, and in this notation the normal components of stress are written $\sigma_{11}, \sigma_{22}, \sigma_{33}$, and the shearing components σ_{23}, σ_{31}, and σ_{12}. In Cartesian coordinates the suffixes 1, 2, and 3 correspond to x, y, and z respectively as used in the text. A general component of stress is referred to simply as σ_{ij} (where i and j can take the values 1, 2, or 3). Similarly the general component of strain is denoted by ϵ_{ij}.

It may be seen that the dilatation in this notation could be expressed as $\sum \epsilon_{ii}$, and if the *summation convention* is used the summation sign is omitted so that ϵ_{ii} is written for $\epsilon_{11} + \epsilon_{22} + \epsilon_{33}$. i is here called a *dummy suffix* and the convention is that any repeated *letter* suffix indicates summation over the values 1, 2, and 3 for that suffix. This convention also applies to derivatives, so that the equation

$$\rho \frac{\partial^2 u_i}{\partial t^2} = \frac{\partial \sigma_{ij}}{\partial x_j} \tag{A.1}$$

corresponds to the three equations

$$\rho\,\frac{\partial^2 u_1}{\partial t^2} = \frac{\partial \sigma_{11}}{\partial x_1} + \frac{\partial \sigma_{12}}{\partial x_2} + \frac{\partial \sigma_{13}}{\partial x_3},$$

$$\rho\,\frac{\partial^2 u_2}{\partial t^2} = \frac{\partial \sigma_{21}}{\partial x_1} + \frac{\partial \sigma_{22}}{\partial x_2} + \frac{\partial \sigma_{23}}{\partial x_3},$$

$$\rho\,\frac{\partial^2 u_3}{\partial t^2} = \frac{\partial \sigma_{31}}{\partial x_1} + \frac{\partial \sigma_{32}}{\partial x_2} + \frac{\partial \sigma_{33}}{\partial x_3},$$

i taking the values 1, 2, and 3 in the first, second, and third equation respectively. These are the same as the three equations of motion of a solid (see equations 2.7) where x_1, x_2, and x_3 are written for the coordinates x, y, and z respectively, and u_1, u_2, and u_3 are written for the components of displacement u, v, and w respectively. The tensor form, equation (A.1), is thus very much more compact than the full Cartesian expressions in (2.7).

Vector form of wave equations

The equations of motion of an elastic solid may be expressed conveniently in vector form, and the vector notation has the advantage of not depending on any one coordinate system, and thus transformations to any given system of coordinates can be effected.

If the displacement is denoted by a vector **s** the components of which are (u, v, w) in a Cartesian system of coordinates (x, y, z), the dilatation Δ, which is a scalar of magnitude $[\partial u/\partial x + \partial v/\partial y + \partial w/\partial z]$, is equal to the *divergence* of **s**, written div **s**. The rotation $\boldsymbol{\omega}$ is a vector the components of which in Cartesian coordinates are $\bar{\omega}_x$, $\bar{\omega}_y$, $\bar{\omega}_z$, and in vector notation this is written as $\frac{1}{2}$ curl **s**. If **i**, **j**, **k** represent unit vectors along the x, y, and z directions respectively, then

$$\text{curl}\,\mathbf{s} = \left(\frac{\partial w}{\partial y} - \frac{\partial v}{\partial z}\right)\mathbf{i} + \left(\frac{\partial u}{\partial z} - \frac{\partial w}{\partial x}\right)\mathbf{j} + \left(\frac{\partial v}{\partial x} - \frac{\partial u}{\partial y}\right)\mathbf{k}.$$

There is one other function which must be defined and that is the *gradient*. This is a vector the components of which are proportional to the rate of change of the derivatives of a scalar quantity. Thus if V is a scalar quantity which is a function of x, y, and z

$$\text{grad}\,V = \frac{\partial V}{\partial x}\mathbf{i} + \frac{\partial V}{\partial y}\mathbf{j} + \frac{\partial V}{\partial z}\mathbf{k}.$$

Now as shown in the textbooks of vector analysis (e.g. Weatherburn, 1924) a number of relations may be established between vector functions. Thus when Laplace's operator ∇^2 operates on a vector **s** so that

$$\nabla^2\mathbf{s} = \mathbf{i}\nabla^2 u + \mathbf{j}\nabla^2 v + \mathbf{k}\nabla^2 w,$$

where $\qquad \nabla^2 u = \left[\dfrac{\partial^2 u}{\partial x^2} + \dfrac{\partial^2 u}{\partial y^2} + \dfrac{\partial^2 u}{\partial z^2}\right], \quad$ etc.,

it may be shown that

$$\nabla^2\mathbf{s} = \text{grad div}\,\mathbf{s} - \text{curl curl}\,\mathbf{s}. \tag{A.2}$$

REFERENCES

ALEXANDROV, A. P., and LAZURKIN, J. S. (1940), *Acta Physicochimica U.R.S.S.*, **12**, 647.

ALFREY, T. (1948), *Mechanical Behaviour of High Polymers* (Interscience: New York).

ALLEGRETTI, L. (1948), *Ric. Sci.*, **18**, 995.

BALLOU, J. W., and SILVERMAN, S. (1944), *J. Acoust. Soc. Amer.*, **16**, 113.

—— and SMITH, J. C. (1949), *J. Appl. Phys.*, **20**, 493.

BANCROFT, D. (1941), *Phys. Rev.*, **59**, 588.

—— and JACOBS, R. B. (1938), *Rev. Sci. Instrum.*, **9**, 279.

BARDUCCI, I., and PASQUALINI, G. (1948), *Nuovo Cimento*, **5**, 416.

BECKER, R., and DÖRING, W. (1939), *Ferromagnetismus* (Springer: Berlin).

BENNEWITZ, K., and RÖTGER, H. (1938), *Zeits. f. techn. Physik*, **19**, 521.

BERGMANN, L. (1949), *Der Ultraschall und seine Anwendung in Wissenschaft und Technik* (Herzel-Verlag: Zürich).

BOLTZMANN, L. (1876), *Pogg. Ann. Erg.*, **7**, 624.

BORDONI, P. G. (1947), *Nuovo Cimento*, **4**, 177.

BRADFIELD, G. (1950), ibid., Supplement No. 2, **7**, 162.

—— (1951), *Nature*, **167**, 1021.

BRIDGMAN, P. W. (1931), *The Physics of High Pressures* (Bell: London).

CHREE, C. (1889 *a*), *Trans. Camb. Phil. Soc.*, **14**, 250.

—— (1889 *b*), *Quart J. Pure and Appl. Math.*, **23**, 335.

CHRISTIE, D. G. (1952) *Trans. Soc. Glass Tech.* **36T**, 74.

COKER, E. G., and FILON, L. N. G. (1931), *Treatise on Photoelasticity* (University Press: Cambridge).

COOKE, W. T. (1936), *Phys. Rev.*, **50**, 1158.

COOPER, J. L. B. (1947), *Phil. Mag.*, **38**, 1.

COURANT, R., and FRIEDRICHS, K. O. (1950), *Supersonic Flow and Shock Waves* (Interscience: New York).

CZERLINSKY, E. VON (1942), *Akust. Z.*, **7**, 12.

DAVIES, R. M. (1948), *Phil. Trans.* A, **240**, 375.

—— (1950), *Some Recent Developments in Rheology*, p. 27 (United Trade Press: London).

—— and JAMES, E. G. (1934), *Phil. Mag.*, **18**, 1023.

—— and OWEN, J. D. (1950), *Proc. Roy. Soc.* A, **204**, 17.

DEBYE, P., and SEARS, F. W. (1932), *Proc. Nat. Acad. Sci. Wash.*, **18**, 410.

DEJUHASZ, K. J. (1949), *J. Franklin Inst.*, **248**, 113.

EDGERTON, H. E., and BARSTOW, F. E. (1939), *J. Am. Ceram. Soc.*, **22**, 302.

—— —— (1941), ibid., **24**, 131.

FIELD, G. S. (1931), *Canad. J. Research*, **5**, 619.

Consider a vector function **F** which has components F_1, F_2, and F_3 in the direction of the coordinate axes in such a system of *orthogonal curvilinear* coordinates, and let **a**, **b**, and **c** be unit vectors along these three directions so that

$$\mathbf{F} = F_1\,\mathbf{a} + F_2\,\mathbf{b} + F_3\,\mathbf{c}.$$

The various functions of this vector may then be expressed in terms of these components and h_1, h_2, and h_3 (see Weatherburn, 1924); for example the divergence is given by

$$\operatorname{div}\mathbf{F} = \frac{1}{h_1 h_2 h_3}\left[\frac{\partial}{\partial q_1}(h_2 h_3 F_1) + \frac{\partial}{\partial q_2}(h_1 h_3 F_2) + \frac{\partial}{\partial q_3}(h_1 h_2 F_3)\right], \quad (A.7)$$

also

$$\operatorname{curl}\mathbf{F} = \frac{\mathbf{a}}{h_2 h_3}\left[\frac{\partial}{\partial q_2}(h_3 F_3) - \frac{\partial}{\partial q_3}(h_2 F_2)\right] +$$

$$+ \frac{\mathbf{b}}{h_3 h_1}\left[\frac{\partial}{\partial q_3}(h_1 F_1) - \frac{\partial}{\partial q_1}(h_3 F_3)\right] +$$

$$+ \frac{\mathbf{c}}{h_1 h_2}\left[\frac{\partial}{\partial q_1}(h_2 F_2) - \frac{\partial}{\partial q_2}(h_1 F_1)\right]. \quad (A.8)$$

Finally, for the gradient of a scalar function V we have

$$\operatorname{grad} V = \frac{\mathbf{a}}{h_1}\frac{\partial V}{\partial q_1} + \frac{\mathbf{b}}{h_2}\frac{\partial V}{\partial q_2} + \frac{\mathbf{c}}{h_3}\frac{\partial V}{\partial q_3}. \quad (A.9)$$

In order to transform (A.7), (A.8), and (A.9) into any particular coordinate system we must find the values for h_1, h_2, and h_3 for that system, and we shall here derive the expressions for a system of cylindrical coordinates r, θ, and z, since these were used in Chapter III. For cylindrical coordinates $dl_1 = dr$, $dl_2 = r\,d\theta$, and $dl_3 = dz$, so that $h_1 = 1$, $h_2 = r$, and $h_3 = 1$. Thus if **s** is the displacement vector and its components are u_r, u_θ, and u_z in the r, θ, and z directions respectively, the dilatation div **s** is from (A.7) given by

$$\Delta = \frac{1}{r}\left[\frac{\partial(r u_r)}{\partial r} + \frac{\partial u_\theta}{\partial \theta} + \frac{\partial(r u_z)}{\partial z}\right]$$

$$= \frac{1}{r}\frac{\partial(r u_r)}{\partial r} + \frac{1}{r}\frac{\partial u_\theta}{\partial \theta} + \frac{\partial u_z}{\partial z}.$$

This is the equation given as (3.38) in Chapter III. Similarly from (A.8) the three components of curl **s** may be obtained and these give the relations (3.39) for $\bar{\omega}_r$, $\bar{\omega}_\theta$, and $\bar{\omega}_z$.

The equation of motion of an isotropic elastic solid is given in vector form by equation (A.5). This may be transformed into cylindrical coordinates by the use of relations (A.8) and (A.9) and this gives equations (3.35), (3.36), and (3.37), which are the Pochhammer relations used in Chapter III for studying the propagation of elastic waves along cylindrical bars. Similar relations may be obtained in spherical coordinates (r, θ, ϕ) and in this case $h_1 = 1$, $h_2 = r$, and $h_3 = r\sin\theta$.

TABLE OF NUMERICAL VALUES

Approximate Numerical Values of the Velocities of Elastic Waves in Some Solids

	Steel	*Copper*	*Aluminium*	*Glass*	*Rubber*
Elastic constants (dynes/sq. cm.)					
λ	$11 \cdot 2 \ \times 10^{11}$	$9 \cdot 5 \ \times 10^{11}$	$5 \cdot 6 \ \times 10^{11}$	$2 \cdot 8 \ \times 10^{11}$	$1 \cdot 0 \ \times 10^{10}$
μ	$8 \cdot 1$,,	$4 \cdot 5$,,	$2 \cdot 6$,,	$2 \cdot 8$,,	$7 \cdot 0 \ \times 10^{6}$
E	$21 \cdot 0$,,	$12 \cdot 0$,,	$7 \cdot 0$,,	$7 \cdot 0$,,	$2 \cdot 0 \ \times 10^{7}$
k	$16 \cdot 7$,,	$12 \cdot 5$,,	$7 \cdot 3$,,	$4 \cdot 7$,,	$1 \cdot 0 \ \times 10^{10}$
Poisson's ratio ν	$0 \cdot 29$	$0 \cdot 34$	$0 \cdot 34$	$0 \cdot 25$	$0 \cdot 5$
Density ρ	$7 \cdot 8$	$8 \cdot 9$	$2 \cdot 7$	$2 \cdot 5$	$0 \cdot 93$
Velocities (metres/sec.)					
c_1	5,940	4,560	6,320	5,800	1,040
c_2	3,220	2,250	3,100	3,350	27
c_0	5,190	3,670	5,090	5,300	46
c_S	2,980	2,120	2,920	3,080	26
c_L	5,420	3,900	5,410	5,460	53

Now the equations of motion of an isotropic elastic solid were derived in Chapter II and are given as equations (2.8), (2.9), and (2.10). These can be written as a single vector equation

$$\rho \frac{\partial^2 \mathbf{s}}{\partial t^2} = (\lambda + \mu) \text{grad div } \mathbf{s} + \mu \nabla^2 \mathbf{s}. \tag{A.3}$$

Substituting for $\nabla^2 \mathbf{s}$ from (A.2) this becomes

$$\rho \frac{\partial^2 \mathbf{s}}{\partial t^2} = (\lambda + 2\mu) \text{grad div } \mathbf{s} - \mu \text{ curl curl } \mathbf{s}. \tag{A.4}$$

If we write (A.4) in terms of the scalar quantity $\Delta \; (= \text{div } \mathbf{s})$ and the vector quantity $\boldsymbol{\omega} \; (= \frac{1}{2} \text{curl } \mathbf{s})$ it becomes

$$\rho \frac{\partial^2 \mathbf{s}}{\partial t^2} = (\lambda + 2\mu) \text{grad } \Delta - 2\mu \text{ curl } \boldsymbol{\omega}. \tag{A.5}$$

These equations are independent of the system of coordinates used as well as being in a more compact form than their Cartesian equivalents.

Curvilinear orthogonal coordinates

Whilst Cartesian coordinates are convenient for many problems in the theory of elasticity, it is often necessary to transform the equations to some other coordinate system, since in some problems the Cartesian expressions are unnecessarily cumbersome. An example of this was found in Chapter III in considering the propagation of elastic waves along cylindrical bars, where cylindrical coordinates were employed.

In general if (q_1, q_2, q_3) are the coordinates of a point in any system and (x, y, z) are its Cartesian coordinates, x, y, and z will be functions of q_1, q_2, and q_3, so that $x = x(q_1, q_2, q_3)$, $y = y(q_1, q_2, q_3)$, and $z = z(q_1, q_2, q_3)$, so that

$$dx = \frac{\partial x}{\partial q_1} dq_1 + \frac{\partial x}{\partial q_2} dq_2 + \frac{\partial x}{\partial q_3} dq_3, \tag{A.6}$$

with similar expressions for y and z.

An element of length $dl^2 = dx^2 + dy^2 + dz^2$ so that from (A.6) we have

$$dl^2 = g_{11} dq_1^2 + g_{22} dq_2^2 + g_{33} dq_3^2 + 2g_{23} dq_2 dq_3 + 2g_{31} dq_3 dq_1 + 2g_{12} dq_1 dq_2,$$

where

$$g_{ij} = g_{ji} = \frac{\partial x}{\partial q_i} \frac{\partial x}{\partial q_j} + \frac{\partial y}{\partial q_i} \frac{\partial y}{\partial q_j} + \frac{\partial z}{\partial q_i} \frac{\partial z}{\partial q_j}.$$

If we consider the length dl_1 which corresponds to a change from q_1 to $q_1 + dq_1$ when q_2 and q_3 are constant, we have $dl_1 = (g_{11})^{\frac{1}{2}} dq_1$. With similar notation $dl_2 = (g_{22})^{\frac{1}{2}} dq_2$ and $dl_3 = (g_{33})^{\frac{1}{2}} dq_3$. For convenience h_1, h_2, and h_3 are written for $(g_{11})^{\frac{1}{2}}$, $(g_{22})^{\frac{1}{2}}$, and $(g_{33})^{\frac{1}{2}}$ respectively. When the system of coordinates is orthogonal, i.e. when the surfaces of $q_1 = \text{constant}$, $q_2 = \text{constant}$, $q_3 = \text{constant}$ always cut each other at right angles, an element of volume dV is given by

$$dV = h_1 h_2 h_3 dq_1 dq_2 dq_3.$$

FÖPPL, O. (1936), *J. Iron and Steel Inst.*, **134**, 393.

FRIEDLANDER, F. G. (1948), *Quart. J. Mech. and Appl. Math.*, **1**, 376.

FROCHT, M. M. (1941), *Photoelasticity* (Wiley: New York).

FUKADA, E. (1950), *J. Phys. Soc. Japan*, **5**, 321.

GALT, J. K. (1948), *Phys. Rev.*, **73**, 1460.

GEMANT, A. (1940), *J. Appl. Phys.*, **11**, 647.

—— and JACKSON, W. (1937), *Phil. Mag.*, **23**, 960.

GIEBE, E., and BLECHSCHMIDT, E. (1933), *Ann. d. Phys.*, **18**, 417.

—— and SCHEIBE, A. (1931), ibid., **9**, 93.

GORSKY, W. S. (1936), *Phys. Zeit. Sowjet.*, **6**, 77.

GRIME, G., and EATON, J. E. (1937), *Phil. Mag.*, **23**, 96.

GROSS, B. (1947), *J. Appl. Phys.*, **18**, 212.

GUILLET, L. (1948), Rep. conf. *Strength of Solids*, p. 116 (Physical Society: London).

GUREVICH, L. (1945), *J. Phys. U.S.S.R.*, **9**, 383.

HERPIN, A. (1947), *Rev. Sci. Paris*, **85**, 817.

HIEDEMANN, E., and HOESCH, K. H. (1935), *Naturwiss.* **23**, 577.

—— ASBACH, H. R., and HOESCH, K. H. (1934), *Z. Phys.*, **90**, 322.

HILLIER, K. W. (1949), *Proc. Phys. Soc. B*, **62**, 701.

—— (1951), ibid., **64**, 998.

—— and KOLSKY, H. (1949), ibid., **62**, 111.

HONDA, K., and KONNO, S. (1921), *Phil. Mag.*, **42**, 115.

HOPKINSON, B. (1905), *Proc. Roy. Soc. A*, **74**, 498.

—— (1910), *Collected Scientific Papers*, p. 64 (University Press: Cambridge, 1921).

—— (1912), ibid., p. 423 (University Press: Cambridge, 1921).

—— (1914), *Phil. Trans. A*, **213**, 437.

—— and WILLIAMS, G. T. (1912), *Proc. Roy. Soc. A*, **87**, 502.

HOPKINSON, J. (1872), *Collected Scientific Papers*, vol. ii, p. 316.

HUDSON, G. E. (1943), *Phys. Rev.*, **63**, 46.

HUGHES, D. S., PONDROM, W. L., and MIMS, R. L. (1949), ibid., **75**, 1552.

HUNTER, L., and SIEGEL, S. (1942), ibid., **61**, 84.

HUNTINGDON, H. B. (1947), ibid., **72**, 321.

IVEY, D. G., MROWCA, B. A., and GUTH, E. (1949), *J. Appl. Phys.*, **20**, 486.

JAMES, E. G., and DAVIES, R. M. (1934), *Phil. Mag.*, **18**, 1053.

JESSOP, H. T., and HARRIS, F. C. (1949), *Photoelasticity* (Cleaver-Hume: London) (Dover reprint).

KAMEL, R. (1949), *Phys. Rev.* **75**, 1606.

KÊ, T. S. (1947), ibid., **71**, 533 and **72**, 41.

KELVIN (1904), Baltimore Lectures (London).

KIMBALL, A. L. (1941), *J. Appl. Mech.*, **8**, A 37 and A 135.

KNOTT, C. G. (1899), *Phil. Mag.*, **48**, 64.

KOLSKY, H. (1949), *Proc. Phys. Soc. B*, **62**, 676.

—— and CHRISTIE, D. G. (1952), *Trans. Soc. Glass Tech.* **36**T, 88.

KOLSKY, H., and SHEARMAN, A. C. (1949), *Research*, **2**, 384.

KOVACS, A. (1948), *C. R. Acad. Sci. Paris*, **227**, 1019.

KRUGER, F., and ROHLOFF, E. (1938), *Zeits. f. Phys.*, **110**, 1, 58.

LAMB, H. (1917), *Proc. Roy. Soc. A*, **93**, 114.

—— (1932), *Hydrodynamics* (Dover reprint).

LANDAU, L., and RUMER, G. (1937), *Phys. Zeit. Sowjet.*, **11**, 18.

LANDON, J. W., and QUINNEY, H. (1923), *Proc. Roy. Soc. A*, **103**, 622.

LE ROLLAND, P. (1931), *C. R. Acad. Sci. Paris*, **192**, 336.

—— (1948), ibid., **227**, 37.

LEADERMAN, H. (1943), *Elastic and Creep Properties of Filamentous Materials* (Textile Foundation: Washington).

LETHERSICH, W. (1946), *Proc. 7th Int. Congress Appl. Mech. London*, Paper I. 6.

—— (1950), *Brit. J. Appl. Phys.*, **1**, 294.

LOVE, A. E. H. (1911), *Some Problems of Geodynamics*, pp. 89–104 and 149–52 (University Press: Cambridge).

—— (1927), *The Mathematical Theory of Elasticity*, 4th edition (Dover reprint).

LUCAS, R., and BIQUARD, P. (1932), *C. R. Acad. Sci. Paris*, **194**, 2132.

MACELWANE, J. B., and SOHON, F. W. (1936), *Introduction to Theoretical Seismology* (Wiley: New York).

MASON, W. P. (1947), *Trans. A.S.M.E.*, **69**, 359.

—— and McSKIMIN, H. J. (1947), *J. Acoust. Soc. Amer.*, **19**, 464.

—— —— (1948), *J. Appl. Phys.*, **19**, 940.

MAXWELL, C. (1890), *Scientific Papers*, vol. ii, p. 26 (University Press: Cambridge).

MEYER, O. (1874), *J. reine angew. Math.*, **58**, 130.

MINDLIN, J. A. (1946), *C. R. Doklady U.R.S.S.*, **1**, 11.

MORSE, R. W. (1948), *J. Acoust. Soc. Amer.*, **20**, 833.

—— (1950), ibid., **22**, 219.

NAVIER, C. L. M. H. (1827), *Mem. Acad. Sci. Paris*, t. 7.

NOLLE, A. W. (1948), *J. Appl. Phys.*, **19**, 753.

—— (1949), *J. Polymer Sci.*, **5**, 1.

NORTON, J. T. (1939), *Rev. Sci. Instrum.*, **10**, 77.

NURMI, U. (1941), *Soc. Sci. fenn. Comm. phys. math.*, **11**, 1.

OROWAN, E. (1934), *Zeits. f. Phys.*, **89**, 634.

OWEN, J. D. (1950), Thesis submitted to University of Wales for Ph.D. degree.

—— and DAVIES, R. M. (1949), *Nature*, **164**, 752.

PARFITT, G. G. (1949), ibid., **164**, 489.

PENNEY, W. G., and PIKE, H. H. (1950), *Rep. Progress Phys.*, **13**, 46 (Physical Society: London).

PERRY, J. (1906), *Phil. Mag.*, **11**, 107.

PLANCK, M. (1932), *Mechanics of Deformable Bodies* (Macmillan: London).

POCHHAMMER, L. (1876), *J. reine angew. Math.*, **81**, 324.

POISSON, S. D. (1829), *Mem. Acad. Sci. Paris*, t. 8.

POLANYI, M. (1934), *Zeits. f. Phys.*, **89**, 660.

PRESCOTT, J. (1942), *Phil. Mag.*, **33**, 703.

QUIMBY, S. L. (1925), *Phys. Rev.*, **25**, 558.

—— (1932), ibid., **39**, 345.

RAKHMATULIN, K. A. (1945), *Appl. Math. and Mech.*, **9**, No. 1.

RANDALL, R. H., ROSE, F. C., and ZENER, C. (1939), *Phys. Rev.*, **56**, 343.

RAYLEIGH (1887), *London Math. Soc. Proc.*, **17**.

—— (1894), *Theory of Sound* (Dover reprint).

RINEHART, J. S. (1951), *J. Appl. Phys.*, **22**, 1178.

ROBERTSON, R. (1921), *Trans. Chem. Soc.*, **119**, 1.

RÖHRICH, K. (1932), *Zeits. f. Phys.*, **73**, 813.

ROTH, W. (1948), *J. Appl. Phys.*, **19**, 901.

SCHAEFER, C., and BERGMANN, L. (1934), *Sitz.-Ber. Berliner Akad.*, pp. 155, 192.

SCHARDIN, H. (1950), *Glastechn. Ber.*, **23**, 1, 67, 325.

—— and STRUTH, W. (1938), ibid., **16**, 219.

SCHOENECK, H. (1934), *Zeits. f. Phys.*, **92**, 390.

SENIOR, D. A., and WELLS, A. A. (1946), *Phil. Mag.*, **37**, 463.

SHEAR, S. K., and FOCKE, A. B. (1940), *Phys. Rev.*, **57**, 532.

SNOEK, J. L. (1941), *Physica*, **8**, 711.

SOUTHWELL, R. V. (1941), *Theory of Elasticity* (Clarendon Press: Oxford).

STANFORD, E. G. (1950), *Nuovo Cimento*, Supplement No. 2, **7**, 332.

—— (1950), Thesis submitted to University of London for Ph.D. degree.

STEPHENS, R. W. B., and BATE, A. E. (1950), *Wave Motion and Sound* (Arnold: London).

STOKES, G. G. (1848), *Phil. Mag.*, **32**, 343.

STONELEY, R. (1924), *Proc. Roy. Soc.* A, **106**, 416.

TAYLOR, G. I. (1934), ibid., **145**, 362.

—— (1946), *J. Inst. Civil Engrs.*, **26**, 486 (James Forrest Lecture).

—— and MACCOLL, J. W. (1935), *The Mechanics of Compressible Fluids*, ed. W. F. Durand, vol. iii (Springer: Berlin).

THIEDE, H. (1941), *Akust. Z.*, **6**, 64.

THOMPSON, J. H. C. (1933), *Phil. Trans.*, **231**, 339.

TIMOSHENKO, S. (1921), *Phil. Mag.*, **41**, 744.

—— (1934), *Theory of Elasticity* (McGraw-Hill: New York).

TOBOLSKY, A., POWELL, R. E., and EYRING, H. (1943), *The Chemistry of Large Molecules*, p. 125 (Interscience: New York).

TRELOAR, L. R. G. (1949), *The Physics of Rubber Elasticity*, pp. 211–33 (Clarendon Press: Oxford).

VAN DYKE, K. S. (1938), *Phys. Rev.*, **53**, 945.

VOIGT, W. (1892), *Ann. d. phys.*, **47**, 671.

VOLTERRA, E. (1948), *Nuovo Cimento*, **4**, 1.

VOLTERRA, V. (1931), *Theory of Functionals* (Dover reprint).

VON KÁRMÁN, T. (1910), 'Festigkeitsprobleme in Maschinenbau', *Ency. d. math. Wiss.*, **4**, Art. 27.

—— and DUWEZ, P. (1950), *J. Appl. Phys.*, **21**, 987.

WEATHERBURN, C. E. (1924), *Advanced Vector Analysis* (Bell: London).

WEBER, W. (1837), *Pogg. Ann., Erg.*, **34**, 247.

WEGEL, R. L., and WALTHER, H. (1935), *Physics*, **6**, 141.

WERT, C. A. (1949), *J. Appl. Phys.*, **20**, 29.

WHITE, M. P. (1949), *J. Appl. Mech.*, **16**, 39.

WHITTAKER, E. J., and WATSON, G. N. (1946), *A Course of Modern Analysis*, p. 359 (University Press: Cambridge).

WITTE, R. S., MROWCA, B. A., and GUTH, E. (1949), *J. Appl. Phys.*, **20**, 481.

ZACHARIAS, J. (1933), *Phys. Rev.*, **44**, 116.

ZENER, C. (1937), ibid., **52**, 230.

—— (1948), *Elasticity and Anelasticity of Metals* (University Press: Chicago).

—— (1949), *Physica*, **15**, 111.

—— and Randall, R. H. (1940), *Trans. A.I.M.E.*, **137**, 41.

ZOEPPRITZ, K. (1919), *Nach. d. königl. Gesell. d. Wissen. z. Göttingen, math.-phys. Berlin*, pp. 66–84.

LIST OF BOOKS AND REVIEW ARTICLES ON SUBJECTS DISCUSSED IN THIS MONOGRAPH WHICH HAVE APPEARED SINCE 1953.

1. DAVIES, R. M. (1953), 'Stress waves in solids', *Appl. Mech. Rev.*, **6**, 1–3.

2. RINEHART, J. S., and PEARSON, J. (1954), *Behavior of Metals under Impulsive Loads* (Am. Soc. for Metals: Cleveland).

3. DAVIES, R. M. (1956), 'Stress waves in solids', *Brit. J. Appl. Phys.*, **7**, 203–09.

4. BROBERG, K. B. (1956), 'Shock waves in elastic and elastic-plastic media', *Kungl. Fortif. Befast. Forsk. och Forsok.*, Report No. 109, 12 (Stockholm).

5. DAVIES, R. M. (1956), 'Stress waves in solids', *Surveys in Mechanics* (G. I. Taylor 70th Anniversary Vol.), pp. 64–138 (University Press: Cambridge).

6. LEE, E. H. (1956), 'Wave propagation in anelastic materials', *Deformation and Flow of Solids* (ed. by R. Grammel), pp. 129–36 (Springer: Berlin).

7. EWING, W. M., JARDETSKY, W. S., and PRESS, F. (1957), *Elastic Waves in Layered Media* (McGraw-Hill: New York).

8. *Proceedings of the Conference on the Properties of Materials at High Rate of Strain* (1957), (Inst. Mech. Eng.: London).

9. ABRAMSON, H. N., PLASS,H. J., and RIPPERGER, E. A. (1958), 'Stress wave propagation in rods and beams', *Advances in Appl. Mech.*, vol: 5, pp. 111–94.

10. KOLSKY, H. (1958), 'The propagation of stress waves in viscoelastic solids', *Appl. Mech. Rev.*, **11**, 465–68.

11. HUNTER, S. C. (1959), 'Viscoelastic waves', *Progress in Solid Mechanics* (ed. by I. N. Sneddon and R. Hill), vol. i, pp. 1–57 (North-Holland Publishing Company: Amsterdam).

12. GREEN, W. A. (1959), 'Dispersion relations for elastic waves in bars', *Progress in Solid Mechanics* (ed. by I. N. Sneddon and R. Hill), vol. i, pp. 225–61 (North–Holland Publishing Company: Amsterdam).

13. KOLSKY, H. (1959), 'Fractures produced by stress waves', *Fracture* (ed. by B. L. Averbach, D. K. Felbeck, G. T. Hahn, and D. A. Thomas), pp. 281–96 (Technology Press and Wiley: New York).

14. *International Symposium on Stress Wave Propagation in Materials* (ed. by N. Davids) (1960), (Interscience: New York).

15. MINDLIN, R. D. (1960), 'Waves and vibrations in isotropic elastic plates', *Structural Mechanics* (ed. by J. N. Goodier and N. J. Hoff), pp. 199–232 (Pergamon: New York).

16. KOLSKY, H. (1960), 'Experimental wave-propagation in solids', *Structural Mechanics* (ed. by J. N. Goodier and N. J. Hoff), pp. 233–62 (Pergamon: New York).

17. CRISTESCU, N. (1960), 'European contributions to dynamic loading and plastic waves', *Plasticity* (ed. by E. H. Lee and P. S. Symonds), pp. 385–442 (Pergamon: New York).

18. GOLDSMITH, W. (1960), *Impact* (Edward Arnold: London).

19. MIKLOWITZ, J. (1960), 'Recent developments in elastic wave propagation', *Appl. Mech. Rev.*, **13**, 865–78.

20. BLAND, D. R. (1960), *The Theory of Linear Viscoelasticity* (Pergamon: London).

21. HOPKINS, H. G. (1961), 'Dynamic anelastic deformations of metals', *Appl. Mech. Rev.*, **14**, 417–31.

22. MUSGRAVE, M. J. P. (1961), 'Elastic waves in anisotropic media', *Progress in Solid Mechanics* (ed. by I. N. Sneddon and R. Hill), vol. ii, pp. 63–85 (North–Holland Publishing Company: Amsterdam).

23. CRAGGS, J. W. (1961), 'Plastic waves', *Progress in Solid Mechanics* (ed. by I. N. Sneddon and R. Hill), vol. ii, pp. 143–97 (North-Holland Publishing Company: Amsterdam).

24. HILLIER, K. W. (1961), 'The measurement of dynamic elastic properties', *Progress in Solid Mechanics* (ed. by I. N. Sneddon and R. Hill), vol. ii, pp. 201–43 (North–Holland Publishing Company: Amsterdam).

25. DUVALL, G. E. (1962), 'Shock waves in the study of solids', *Appl. Mech. Rev.*, **15**, 849–54.

26. KOLSKY, H. (1962), 'The detection and measurement of stress waves', *Experimental Techniques in Shock and Vibration* (ed. by W. J. Worley), pp. 11–24 (Am. Soc Mech. Eng.).

27. *Stress Waves in Anelastic Solids*, Proceedings of IUTAM Symposium held at Brown University 1963 (Springer: Berlin; in press).

AUTHOR INDEX

SUBJECT INDEX

Acoustic ohms, 44.
Aeolotropic solids: stress-strain relations for, 8; elastic waves in, 38–40; visco-elastic behaviour of, 107.
Attenuation coefficient, 105, 118–21, 143.

Ballistic pendulum measurements, 151–3.
Bars: conical, 75–79, 90; flexural vibrations of, 48–54, 68–73, 85; longitudinal vibrations of, 41–47, 56–65, 85; motion of freely suspended, 45–47; of non-circular cross-section, 74; torsional vibrations of, 47, 65–68, 85.
Boundary conditions: for reflection, 26–31; for refraction, 31–38.
Brewster's law, 148.
Brittle fracture, 195.
Bulk modulus, definition, 9.
Bulk wave, *see* Dilatation wave.

Characteristic impedance, 34, 44.
Components: of displacement, 5; of strain, 6; of stress, 5.
Condenser microphone, 92, 94.
Cones, fractures in, 193.
Conical bars, 75–79, 90.
Crack propagation, velocity of, 183, 194.
Critical angle, 36.
Crystals: single, *see* Aeolotropic solids; dynamic measurements with, 96, 131.
Curvilinear coordinates, 199.
Cylinders, *see* Bars.

Davies bar, 91–95, 153–6.
Dead zone, 65.
Debye modes, 128.
Dilatation, expressions for, 9, 55, 84, 198.
Dilatation waves, 4, 12 ff.; in a bar, 63; reflection of, 24–29, 32–34.
Dilatational viscosity, 109.
Dislocations, 128.
Dispersion: in bars, 47, 49, 51, 62, 67, 73, 94; in plates, 83; of pulses, 73, 146; in visco-elastic media, 120–2.

Displacement: components of, 5; at end of free bar, 44.
Distortion wave, 4, 12 ff.; reflection of, 29–31, 34–36.
Ductile failure, 195.

Elastic constants: for aeolotropic solids, 8; for isotropic solids, 9; measurement of, 49, 130–62; numerical values, 201.
Electrostatic transducer, 138.
Energy flux, 28.
Equations of motion, 10, 198; for aeolotropic solids, 39; for isotropic solids, 12, 54, 84; for plastic solids, 164, 168; for visco-elastic solids, 116–21; vector form of, 199.
Eulerian coordinates, 163, 167.
Extensional viscosity, 118.

Ferromagnetic materials, mechanical losses in, 129.
Filaments, wave propagation in, 142–4.
Flexural waves in bars, 48–54, 68–73, 85.
Fluids, waves in, 1.
Fractures, 183–96; brittle, 195; in cones, 193; in cylinders, 191; by explosives, 186–96; in steel wires, 185; produced by stress reinforcement, 189–93.
Free vibration measurements, 30, 136.
Fresnel surfaces, 40.
Frequency equation: in bars, 57, 60, 70; in plates, 81.
Functionals, theory of, 116.

Glass: dynamic measurements of, 96, 131, 137; fractures in, 189.
Group velocity, 51, 59, 72, 83.

Half-breadth of resonance peak, 103.
High-speed photography, 149.
Hooke's law, 7–9.
Hopkinson bar, 87–91, 188.
Huygens's principle, 32.
Hysteresis, 122.

A CATALOG OF SELECTED
DOVER BOOKS
IN SCIENCE AND MATHEMATICS

A CATALOG OF SELECTED
DOVER BOOKS
IN SCIENCE AND MATHEMATICS

QUALITATIVE THEORY OF DIFFERENTIAL EQUATIONS, V.V. Nemytskii and V.V. Stepanov. Classic graduate-level text by two prominent Soviet mathematicians covers classical differential equations as well as topological dynamics and erqodic theory. Bibliographies. 523pp. 5⅜ × 8½.   65954-2 Pa. $10.95

MATRICES AND LINEAR ALGEBRA, Hans Schneider and George Phillip Barker. Basic textbook covers theory of matrices and its applications to systems of linear equations and related topics such as determinants, eigenvalues and differential equations. Numerous exercises. 432pp. 5⅜ × 8½.   66014-1 Pa. $8.95

QUANTUM THEORY, David Bohm. This advanced undergraduate-level text presents the quantum theory in terms of qualitative and imaginative concepts, followed by specific applications worked out in mathematical detail. Preface. Index. 655pp. 5⅜ × 8½.   65969-0 Pa. $10.95

ATOMIC PHYSICS (8th edition), Max Born. Nobel laureate's lucid treatment of kinetic theory of gases, elementary particles, nuclear atom, wave-corpuscles, atomic structure and spectral lines, much more. Over 40 appendices, bibliography. 495pp. 5⅜ × 8½.   65984-4 Pa. $11.95

ELECTRONIC STRUCTURE AND THE PROPERTIES OF SOLIDS: The Physics of the Chemical Bond, Walter A. Harrison. Innovative text offers basic understanding of the electronic structure of covalent and ionic solids, simple metals, transition metals and their compounds. Problems. 1980 edition. 582pp. 6⅛ × 9¼.   66021-4 Pa. $14.95

BOUNDARY VALUE PROBLEMS OF HEAT CONDUCTION, M. Necati Özisik. Systematic, comprehensive treatment of modern mathematical methods of solving problems in heat conduction and diffusion. Numerous examples and problems. Selected references. Appendices. 505pp. 5⅜ × 8½.   65990-9 Pa. $11.95

A SHORT HISTORY OF CHEMISTRY (3rd edition), J.R. Partington. Classic exposition explores origins of chemistry, alchemy, early medical chemistry, nature of atmosphere, theory of valency, laws and structure of atomic theory, much more. 428pp. 5⅜ × 8½. (Available in U.S. only)   65977-1 Pa. $10.95

A HISTORY OF ASTRONOMY, A. Pannekoek. Well-balanced, carefully reasoned study covers such topics as Ptolemaic theory, work of Copernicus, Kepler, Newton, Eddington's work on stars, much more. Illustrated. References. 521pp. 5⅜ × 8½.   65994-1 Pa. $11.95

PRINCIPLES OF METEOROLOGICAL ANALYSIS, Walter J. Saucier. Highly respected, abundantly illustrated classic reviews atmospheric variables, hydrostatics, static stability, various analyses (scalar, cross-section, isobaric, isentropic, more). For intermediate meteorology students. 454pp. 6⅛ × 9¼. 65979-8 Pa. $12.95

RELATIVITY, THERMODYNAMICS AND COSMOLOGY, Richard C. Tolman. Landmark study extends thermodynamics to special, general relativity; also applications of relativistic mechanics, thermodynamics to cosmological models. 501pp. 5⅜ × 8½. 65383-8 Pa. $11.95

APPLIED ANALYSIS, Cornelius Lanczos. Classic work on analysis and design of finite processes for approximating solution of analytical problems. Algebraic equations, matrices, harmonic analysis, quadrature methods, much more. 559pp. 5⅜ × 8½. 65656-X Pa. $11.95

SPECIAL RELATIVITY FOR PHYSICISTS, G. Stephenson and C.W. Kilmister. Concise elegant account for nonspecialists. Lorentz transformation, optical and dynamical applications, more. Bibliography. 108pp. 5⅜ × 8½. 65519-9 Pa. $3.95

INTRODUCTION TO ANALYSIS, Maxwell Rosenlicht. Unusually clear, accessible coverage of set theory, real number system, metric spaces, continuous functions, Riemann integration, multiple integrals, more. Wide range of problems. Undergraduate level. Bibliography. 254pp. 5⅜ × 8½. 65038-3 Pa. $7.00

INTRODUCTION TO QUANTUM MECHANICS With Applications to Chemistry, Linus Pauling & E. Bright Wilson, Jr. Classic undergraduate text by Nobel Prize winner applies quantum mechanics to chemical and physical problems. Numerous tables and figures enhance the text. Chapter bibliographies. Appendices. Index. 468pp. 5⅜ × 8½. 64871-0 Pa. $9.95

ASYMPTOTIC EXPANSIONS OF INTEGRALS, Norman Bleistein & Richard A. Handelsman. Best introduction to important field with applications in a variety of scientific disciplines. New preface. Problems. Diagrams. Tables. Bibliography. Index. 448pp. 5⅜ × 8½. 65082-0 Pa. $10.95

MATHEMATICS APPLIED TO CONTINUUM MECHANICS, Lee A. Segel. Analyzes models of fluid flow and solid deformation. For upper-level math, science and engineering students. 608pp. 5⅜ × 8½. 65369-2 Pa. $12.95

ELEMENTS OF REAL ANALYSIS, David A. Sprecher. Classic text covers fundamental concepts, real number system, point sets, functions of a real variable, Fourier series, much more. Over 500 exercises. 352pp. 5⅜ × 8½. 65385-4 Pa. $8.95

PHYSICAL PRINCIPLES OF THE QUANTUM THEORY, Werner Heisenberg. Nobel Laureate discusses quantum theory, uncertainty, wave mechanics, work of Dirac, Schroedinger, Compton, Wilson, Einstein, etc. 184pp. 5⅜ × 8½. 60113-7 Pa. $4.95

INTRODUCTORY REAL ANALYSIS, A.N. Kolmogorov, S.V. Fomin. Translated by Richard A. Silverman. Self-contained, evenly paced introduction to real and functional analysis. Some 350 problems. 403pp. 5⅜ × 8½. 61226-0 Pa. $7.95

PROBLEMS AND SOLUTIONS IN QUANTUM CHEMISTRY AND PHYSICS, Charles S. Johnson, Jr. and Lee G. Pedersen. Unusually varied problems, detailed solutions in coverage of quantum mechanics, wave mechanics, angular momentum, molecular spectroscopy, scattering theory, more. 280 problems plus 139 supplementary exercises. 430pp. 6½ × 9¼. 65236-X Pa. $10.95

ASYMPTOTIC METHODS IN ANALYSIS, N.G. de Bruijn. An inexpensive, comprehensive guide to asymptotic methods—the pioneering work that teaches by explaining worked examples in detail. Index. 224pp. 5⅜ × 8½.   64221-6 Pa. $5.95

OPTICAL RESONANCE AND TWO-LEVEL ATOMS, L. Allen and J.H. Eberly. Clear, comprehensive introduction to basic principles behind all quantum optical resonance phenomena. 53 illustrations. Preface. Index. 256pp. 5⅜ × 8½.
65533-4 Pa. $6.95

COMPLEX VARIABLES, Francis J. Flanigan. Unusual approach, delaying complex algebra till harmonic functions have been analyzed from real variable viewpoint. Includes problems with answers. 364pp. 5⅜ × 8½.   61388-7 Pa. $7.95

ATOMIC SPECTRA AND ATOMIC STRUCTURE, Gerhard Herzberg. One of best introductions; especially for specialist in other fields. Treatment is physical rather than mathematical. 80 illustrations. 257pp. 5⅜ × 8½.   60115-3 Pa. $4.95

APPLIED COMPLEX VARIABLES, John W. Dettman. Step-by-step coverage of fundamentals of analytic function theory—plus lucid exposition of 5 important applications: Potential Theory; Ordinary Differential Equations; Fourier Transforms; Laplace Transforms; Asymptotic Expansions. 66 figures. Exercises at chapter ends. 512pp. 5⅜ × 8½.   64670-X Pa. $10.95

ULTRASONIC ABSORPTION: An Introduction to the Theory of Sound Absorption and Dispersion in Gases, Liquids and Solids, A.B. Bhatia. Standard reference in the field provides a clear, systematically organized introductory review of fundamental concepts for advanced graduate students, research workers. Numerous diagrams. Bibliography. 440pp. 5⅜ × 8½.   64917-2 Pa. $8.95

UNBOUNDED LINEAR OPERATORS: Theory and Applications, Seymour Goldberg. Classic presents systematic treatment of the theory of unbounded linear operators in normed linear spaces with applications to differential equations. Bibliography. 199pp. 5⅜ × 8½.   64830-3 Pa. $7.00

LIGHT SCATTERING BY SMALL PARTICLES, H.C. van de Hulst. Comprehensive treatment including full range of useful approximation methods for researchers in chemistry, meteorology and astronomy. 44 illustrations. 470pp. 5⅜ × 8½.   64228-3 Pa. $9.95

CONFORMAL MAPPING ON RIEMANN SURFACES, Harvey Cohn. Lucid, insightful book presents ideal coverage of subject. 334 exercises make book perfect for self-study. 55 figures. 352pp. 5⅜ × 8¼.   64025-6 Pa. $8.95

OPTICKS, Sir Isaac Newton. Newton's own experiments with spectroscopy, colors, lenses, reflection, refraction, etc., in language the layman can follow. Foreword by Albert Einstein. 532pp. 5⅜ × 8½.   60205-2 Pa. $8.95

GENERALIZED INTEGRAL TRANSFORMATIONS, A.H. Zemanian. Graduate-level study of recent generalizations of the Laplace, Mellin, Hankel, K. Weierstrass, convolution and other simple transformations. Bibliography. 320pp. 5⅜ × 8½.   65375-7 Pa. $7.95

THE ELECTROMAGNETIC FIELD, Albert Shadowitz. Comprehensive undergraduate text covers basics of electric and magnetic fields, builds up to electromagnetic theory. Also related topics, including relativity. Over 900 problems. 768pp. 5⅜ × 8¼. 65660-8 Pa. $15.95

FOURIER SERIES, Georgi P. Tolstov. Translated by Richard A. Silverman. A valuable addition to the literature on the subject, moving clearly from subject to subject and theorem to theorem. 107 problems, answers. 336pp. 5⅜ × 8½. 63317-9 Pa. $7.95

THEORY OF ELECTROMAGNETIC WAVE PROPAGATION, Charles Herach Papas. Graduate-level study discusses the Maxwell field equations, radiation from wire antennas, the Doppler effect and more. xiii + 244pp. 5⅜ × 8½. 65678-0 Pa. $6.95

DISTRIBUTION THEORY AND TRANSFORM ANALYSIS: An Introduction to Generalized Functions, with Applications, A.H. Zemanian. Provides basics of distribution theory, describes generalized Fourier and Laplace transformations. Numerous problems. 384pp. 5⅜ × 8½. 65479-6 Pa. $8.95

THE PHYSICS OF WAVES, William C. Elmore and Mark A. Heald. Unique overview of classical wave theory. Acoustics, optics, electromagnetic radiation, more. Ideal as classroom text or for self-study. Problems. 477pp. 5⅜ × 8½. 64926-1 Pa. $10.95

CALCULUS OF VARIATIONS WITH APPLICATIONS, George M. Ewing. Applications-oriented introduction to variational theory develops insight and promotes understanding of specialized books, research papers. Suitable for advanced undergraduate/graduate students as primary, supplementary text. 352pp. 5⅜ × 8½. 64856-7 Pa. $8.50

A TREATISE ON ELECTRICITY AND MAGNETISM, James Clerk Maxwell. Important foundation work of modern physics. Brings to final form Maxwell's theory of electromagnetism and rigorously derives his general equations of field theory. 1,084pp. 5⅜ × 8½. 60636-8, 60637-6 Pa., Two-vol. set $19.00

AN INTRODUCTION TO THE CALCULUS OF VARIATIONS, Charles Fox. Graduate-level text covers variations of an integral, isoperimetrical problems, least action, special relativity, approximations, more. References. 279pp. 5⅜ × 8½. 65499-0 Pa. $6.95

HYDRODYNAMIC AND HYDROMAGNETIC STABILITY, S. Chandrasekhar. Lucid examination of the Rayleigh-Benard problem; clear coverage of the theory of instabilities causing convection. 704pp. 5⅜ × 8¼. 64071-X Pa. $12.95

CALCULUS OF VARIATIONS, Robert Weinstock. Basic introduction covering isoperimetric problems, theory of elasticity, quantum mechanics, electrostatics, etc. Exercises throughout. 326pp. 5⅜ × 8½. 63069-2 Pa. $7.95

DYNAMICS OF FLUIDS IN POROUS MEDIA, Jacob Bear. For advanced students of ground water hydrology, soil mechanics and physics, drainage and irrigation engineering and more. 335 illustrations. Exercises, with answers. 784pp. 6⅛ × 9¼. 65675-6 Pa. $19.95

NUMERICAL METHODS FOR SCIENTISTS AND ENGINEERS, Richard Hamming. Classic text stresses frequency approach in coverage of algorithms, polynomial approximation, Fourier approximation, exponential approximation, other topics. Revised and enlarged 2nd edition. 721pp. 5⅜ × 8½.
65241-6 Pa. $14.95

THEORETICAL SOLID STATE PHYSICS, Vol. I: Perfect Lattices in Equilibrium; Vol. II: Non-Equilibrium and Disorder, William Jones and Norman H. March. Monumental reference work covers fundamental theory of equilibrium properties of perfect crystalline solids, non-equilibrium properties, defects and disordered systems. Appendices. Problems. Preface. Diagrams. Index. Bibliography. Total of 1,301pp. 5⅜ × 8½. Two volumes. Vol. I 65015-4 Pa. $12.95
Vol. II 65016-2 Pa. $12.95

OPTIMIZATION THEORY WITH APPLICATIONS, Donald A. Pierre. Broad-spectrum approach to important topic. Classical theory of minima and maxima, calculus of variations, simplex technique and linear programming, more. Many problems, examples. 640pp. 5⅜ × 8½. 65205-X Pa. $12.95

THE MODERN THEORY OF SOLIDS, Frederick Seitz. First inexpensive edition of classic work on theory of ionic crystals, free-electron theory of metals and semiconductors, molecular binding, much more. 736pp. 5⅜ × 8½.
65482-6 Pa. $14.95

ESSAYS ON THE THEORY OF NUMBERS, Richard Dedekind. Two classic essays by great German mathematician: on the theory of irrational numbers; and on transfinite numbers and properties of natural numbers. 115pp. 5⅜ × 8½.
21010-3 Pa. $4.95

THE FUNCTIONS OF MATHEMATICAL PHYSICS, Harry Hochstadt. Comprehensive treatment of orthogonal polynomials, hypergeometric functions, Hill's equation, much more. Bibliography. Index. 322pp. 5⅜ × 8½. 65214-9 Pa. $8.95

NUMBER THEORY AND ITS HISTORY, Oystein Ore. Unusually clear, accessible introduction covers counting, properties of numbers, prime numbers, much more. Bibliography. 380pp. 5⅜ × 8½. 65620-9 Pa. $8.95

THE VARIATIONAL PRINCIPLES OF MECHANICS, Cornelius Lanczos. Graduate level coverage of calculus of variations, equations of motion, relativistic mechanics, more. First inexpensive paperbound edition of classic treatise. Index. Bibliography. 418pp. 5⅜ × 8½. 65067-7 Pa. $10.95

MATHEMATICAL TABLES AND FORMULAS, Robert D. Carmichael and Edwin R. Smith. Logarithms, sines, tangents, trig functions, powers, roots, reciprocals, exponential and hyperbolic functions, formulas and theorems. 269pp. 5⅜ × 8½. 60111-0 Pa. $5.95

THEORETICAL PHYSICS, Georg Joos, with Ira M. Freeman. Classic overview covers essential math, mechanics, electromagnetic theory, thermodynamics, quantum mechanics, nuclear physics, other topics. First paperback edition. xxiii + 885pp. 5⅜ × 8½. 65227-0 Pa. $17.95

HANDBOOK OF MATHEMATICAL FUNCTIONS WITH FORMULAS, GRAPHS, AND MATHEMATICAL TABLES, edited by Milton Abramowitz and Irene A. Stegun. Vast compendium: 29 sets of tables, some to as high as 20 places. 1,046pp. 8 × 10½. 61272-4 Pa. $21.95

MATHEMATICAL METHODS IN PHYSICS AND ENGINEERING, John W. Dettman. Algebraically based approach to vectors, mapping, diffraction, other topics in applied math. Also generalized functions, analytic function theory, more. Exercises. 448pp. 5⅜ × 8¼. 65649-7 Pa. $8.95

A SURVEY OF NUMERICAL MATHEMATICS, David M. Young and Robert Todd Gregory. Broad self-contained coverage of computer-oriented numerical algorithms for solving various types of mathematical problems in linear algebra, ordinary and partial, differential equations, much more. Exercises. Total of 1,248pp. 5⅜ × 8½. Two volumes. Vol. I 65691-8 Pa. $13.95
Vol. II 65692-6 Pa. $13.95

TENSOR ANALYSIS FOR PHYSICISTS, J.A. Schouten. Concise exposition of the mathematical basis of tensor analysis, integrated with well-chosen physical examples of the theory. Exercises. Index. Bibliography. 289pp. 5⅜ × 8½. 65582-2 Pa. $7.95

INTRODUCTION TO NUMERICAL ANALYSIS (2nd Edition), F.B. Hildebrand. Classic, fundamental treatment covers computation, approximation, interpolation, numerical differentiation and integration, other topics. 150 new problems. 669pp. 5⅜ × 8½. 65363-3 Pa. $13.95

INVESTIGATIONS ON THE THEORY OF THE BROWNIAN MOVEMENT, Albert Einstein. Five papers (1905–8) investigating dynamics of Brownian motion and evolving elementary theory. Notes by R. Fürth. 122pp. 5⅜ × 8½. 60304-0 Pa. $3.95

NUMERICAL METHODS FOR SCIENTISTS AND ENGINEERS, Richard Hamming. Classic text stresses frequency approach in coverage of algorithms, polynomial approximation, Fourier approximation, exponential approximation, other topics. Revised and enlarged 2nd edition. 721pp. 5⅜ × 8½. 65241-6 Pa. $14.95

AN INTRODUCTION TO STATISTICAL THERMODYNAMICS, Terrell L. Hill. Excellent basic text offers wide-ranging coverage of quantum statistical mechanics, systems of interacting molecules, quantum statistics, more. 523pp. 5⅜ × 8½. 65242-4 Pa. $10.95

ELEMENTARY DIFFERENTIAL EQUATIONS, William Ted Martin and Eric Reissner. Exceptionally, clear comprehensive introduction at undergraduate level. Nature and origin of differential equations, differential equations of first, second and higher orders. Picard's Theorem, much more. Problems with solutions. 331pp. 5⅜ × 8½. 65024-3 Pa. $8.95

STATISTICAL PHYSICS, Gregory H. Wannier. Classic text combines thermodynamics, statistical mechanics and kinetic theory in one unified presentation of thermal physics. Problems with solutions. Bibliography. 532pp. 5⅜ × 8½. 65401-X Pa. $10.95

ORDINARY DIFFERENTIAL EQUATIONS, Morris Tenenbaum and Harry Pollard. Exhaustive survey of ordinary differential equations for undergraduates in mathematics, engineering, science. Thorough analysis of theorems. Diagrams. Bibliography. Index. 818pp. 5⅜ × 8½.   .64940-7 Pa. $15.95

STATISTICAL MECHANICS: Principles and Applications, Terrell L. Hill. Standard text covers fundamentals of statistical mechanics, applications to fluctuation theory, imperfect gases, distribution functions, more. 448pp. 5⅜ × 8½.   65390-0 Pa. $9.95

ORDINARY DIFFERENTIAL EQUATIONS AND STABILITY THEORY: An Introduction, David A. Sánchez. Brief, modern treatment. Linear equation, stability theory for autonomous and nonautonomous systems, etc. 164pp. 5⅜ × 8¼.   63828-6 Pa. $4.95

THIRTY YEARS THAT SHOOK PHYSICS: The Story of Quantum Theory, George Gamow. Lucid, accessible introduction to influential theory of energy and matter. Careful explanations of Dirac's anti-particles, Bohr's model of the atom, much more. 12 plates. Numerous drawings. 240pp. 5⅜ × 8½.   24895-X Pa. $5.95

ORDINARY DIFFERENTIAL EQUATIONS, I.G. Petrovski. Covers basic concepts, some differential equations and such aspects of the general theory as Euler lines, Arzel's theorem, Peano's existence theorem, Osgood's uniqueness theorem, more. 45 figures. Problems. Bibliography. Index. xi + 232pp. 5⅜ × 8½.   64683-1 Pa. $6.00

GREAT EXPERIMENTS IN PHYSICS: Firsthand Accounts from Galileo to Einstein, edited by Morris H. Shamos. 25 crucial discoveries: Newton's laws of motion, Chadwick's study of the neutron, Hertz on electromagnetic waves, more. Original accounts clearly annotated. 370pp. 5⅜ × 8½.   25346-5 Pa. $8.95

INTRODUCTION TO PARTIAL DIFFERENTIAL EQUATIONS WITH APPLICATIONS, E.C. Zachmanoglou and Dale W. Thoe. Essentials of partial differential equations applied to common problems in engineering and the physical sciences. Problems and answers. 416pp. 5⅜ × 8½.   65251-3 Pa. $9.95

BURNHAM'S CELESTIAL HANDBOOK, Robert Burnham, Jr. Thorough guide to the stars beyond our solar system. Exhaustive treatment. Alphabetical by constellation: Andromeda to Cetus in Vol. 1; Chamaeleon to Orion in Vol. 2; and Pavo to Vulpecula in Vol. 3. Hundreds of illustrations. Index in Vol. 3. 2,000pp. 6⅛ × 9¼.   23567-X, 23568-8, 23673-0 Pa., Three-vol. set $38.85

ASYMPTOTIC EXPANSIONS FOR ORDINARY DIFFERENTIAL EQUATIONS, Wolfgang Wasow. Outstanding text covers asymptotic power series, Jordan's canonical form, turning point problems, singular perturbations, much more. Problems. 384pp. 5⅜ × 8½.   65456-7 Pa. $8.95

AMATEUR ASTRONOMER'S HANDBOOK, J.B. Sidgwick. Timeless, comprehensive coverage of telescopes, mirrors, lenses, mountings, telescope drives, micrometers, spectroscopes, more. 189 illustrations. 576pp. 5⅜ × 8¼.   24034-7 Pa. $8.95

SPECIAL FUNCTIONS, N.N. Lebedev. Translated by Richard Silverman. Famous Russian work treating more important special functions, with applications to specific problems of physics and engineering. 38 figures. 308pp. 5⅜ × 8½.
60624-4 Pa. $6.95

OBSERVATIONAL ASTRONOMY FOR AMATEURS, J.B. Sidgwick. Mine of useful data for observation of sun, moon, planets, asteroids, aurorae, meteors, comets, variables, binaries, etc. 39 illustrations 384pp. 5⅜ × 8¼. (Available in U.S. only)
24033-9 Pa. $5.95

INTEGRAL EQUATIONS, F.G. Tricomi. Authoritative, well-written treatment of extremely useful mathematical tool with wide applications. Volterra Equations, Fredholm Equations, much more. Advanced undergraduate to graduate level. Exercises. Bibliography. 238pp. 5⅜ × 8½.
64828-1 Pa. $6.95

CELESTIAL OBJECTS FOR COMMON TELESCOPES, T.W. Webb. Inestimable aid for locating and identifying nearly 4,000 celestial objects. 77 illustrations. 645pp. 5⅜ × 8½.
20917-2, 20918-0 Pa., Two-vol. set $12.00

MODERN NONLINEAR EQUATIONS, Thomas L. Saaty. Emphasizes practical solution of problems; covers seven types of equations. ". . . a welcome contribution to the existing literature. . . ."—*Math Reviews.* 490pp. 5⅜ × 8½. 64232-1 Pa. $9.95

FUNDAMENTALS OF ASTRODYNAMICS, Roger Bate et al. Modern approach developed by U.S. Air Force Academy. Designed as a first course. Problems, exercises. Numerous illustrations. 455pp. 5⅜ × 8½.
60061-0 Pa. $8.95

INTRODUCTION TO LINEAR ALGEBRA AND DIFFERENTIAL EQUATIONS, John W. Dettman. Excellent text covers complex numbers, determinants, orthonormal bases, Laplace transforms, much more. Exercises with solutions. Undergraduate level. 416pp. 5⅜ × 8½.
65191-6 Pa. $8.95

INCOMPRESSIBLE AERODYNAMICS, edited by Bryan Thwaites. Covers theoretical and experimental treatment of the uniform flow of air and viscous fluids past two-dimensional aerofoils and three-dimensional wings; many other topics. 654pp. 5⅜ × 8½.
65465-6 Pa. $14.95

INTRODUCTION TO DIFFERENCE EQUATIONS, Samuel Goldberg. Exceptionally clear exposition of important discipline with applications to sociology, psychology, economics. Many illustrative examples; over 250 problems. 260pp. 5⅜ × 8½.
65084-7 Pa. $6.95

LAMINAR BOUNDARY LAYERS, edited by L. Rosenhead. Engineering classic covers steady boundary layers in two- and three-dimensional flow, unsteady boundary layers, stability, observational techniques, much more. 708pp. 5⅜ × 8½.
65646-2 Pa. $15.95

LECTURES ON CLASSICAL DIFFERENTIAL GEOMETRY, Second Edition, Dirk J. Struik. Excellent brief introduction covers curves, theory of surfaces, fundamental equations, geometry on a surface, conformal mapping, other topics. Problems. 240pp. 5⅜ × 8½.
65609-8 Pa. $6.95

ROTARY-WING AERODYNAMICS, W.Z. Stepniewski. Clear, concise text covers aerodynamic phenomena of the rotor and offers guidelines for helicopter performance evaluation. Originally prepared for NASA. 537 figures. 640pp. 6⅛ × 9¼.
64647-5 Pa. $14.95

DIFFERENTIAL GEOMETRY, Heinrich W. Guggenheimer. Local differential geometry as an application of advanced calculus and linear algebra. Curvature, transformation groups, surfaces, more. Exercises. 62 figures. 378pp. 5⅜ × 8½.
63433-7 Pa. $7.95

INTRODUCTION TO SPACE DYNAMICS, William Tyrrell Thomson. Comprehensive, classic introduction to space-flight engineering for advanced undergraduate and graduate students. Includes vector algebra, kinematics, transformation of coordinates. Bibliography. Index. 352pp. 5⅜ × 8½.   65113-4 Pa. $8.00

A SURVEY OF MINIMAL SURFACES, Robert Osserman. Up-to-date, in-depth discussion of the field for advanced students. Corrected and enlarged edition covers new developments. Includes numerous problems. 192pp. 5⅜ × 8½.
64998-9 Pa. $8.00

ANALYTICAL MECHANICS OF GEARS, Earle Buckingham. Indispensable reference for modern gear manufacture covers conjugate gear-tooth action, gear-tooth profiles of various gears, many other topics. 263 figures. 102 tables. 546pp. 5⅜ × 8½.   65712-4 Pa. $11.95

SET THEORY AND LOGIC, Robert R. Stoll. Lucid introduction to unified theory of mathematical concepts. Set theory and logic seen as tools for conceptual understanding of real number system. 496pp. 5⅜ × 8¼.   63829-4 Pa. $8.95

A HISTORY OF MECHANICS, René Dugas. Monumental study of mechanical principles from antiquity to quantum mechanics. Contributions of ancient Greeks, Galileo, Leonardo, Kepler, Lagrange, many others. 671pp. 5⅜ × 8½.
65632-2 Pa. $14.95

FAMOUS PROBLEMS OF GEOMETRY AND HOW TO SOLVE THEM, Benjamin Bold. Squaring the circle, trisecting the angle, duplicating the cube: learn their history, why they are impossible to solve, then solve them yourself. 128pp. 5⅜ × 8½.   24297-8 Pa. $3.95

MECHANICAL VIBRATIONS, J.P. Den Hartog. Classic textbook offers lucid explanations and illustrative models, applying theories of vibrations to a variety of practical industrial engineering problems. Numerous figures. 233 problems, solutions. Appendix. Index. Preface. 436pp. 5⅜ × 8½.   64785-4 Pa. $8.95

CURVATURE AND HOMOLOGY, Samuel I. Goldberg. Thorough treatment of specialized branch of differential geometry. Covers Riemannian manifolds, topology of differentiable manifolds, compact Lie groups, other topics. Exercises. 315pp. 5⅜ × 8½.   64314-X Pa. $6.95

HISTORY OF STRENGTH OF MATERIALS, Stephen P. Timoshenko. Excellent historical survey of the strength of materials with many references to the theories of elasticity and structure. 245 figures. 452pp. 5⅜ × 8½.   61187-6 Pa. $9.95

GEOMETRY OF COMPLEX NUMBERS, Hans Schwerdtfeger. Illuminating, widely praised book on analytic geometry of circles, the Moebius transformation, and two-dimensional non-Euclidean geometries. 200pp. 5⅜ × 8¼.
63830-8 Pa. $6.95

MECHANICS, J.P. Den Hartog. A classic introductory text or refresher. Hundreds of applications and design problems illuminate fundamentals of trusses, loaded beams and cables, etc. 334 answered problems. 462pp. 5⅜ × 8½. 60754-2 Pa. $8.95

TOPOLOGY, John G. Hocking and Gail S. Young. Superb one-year course in classical topology. Topological spaces and functions, point-set topology, much more. Examples and problems. Bibliography. Index. 384pp. 5⅜ × 8¼.
65676-4 Pa. $7.95

STRENGTH OF MATERIALS, J.P. Den Hartog. Full, clear treatment of basic material (tension, torsion, bending, etc.) plus advanced material on engineering methods, applications. 350 answered problems. 323pp. 5⅜ × 8½. 60755-0 Pa. $7.50

ELEMENTARY CONCEPTS OF TOPOLOGY, Paul Alexandroff. Elegant, intuitive approach to topology from set-theoretic topology to Betti groups; how concepts of topology are useful in math and physics. 25 figures. 57pp. 5⅜ × 8½.
60747-X Pa. $2.95

ADVANCED STRENGTH OF MATERIALS, J.P. Den Hartog. Superbly written advanced text covers torsion, rotating disks, membrane stresses in shells, much more. Many problems and answers. 388pp. 5⅜ × 8½. 65407-9 Pa. $8.95

COMPUTABILITY AND UNSOLVABILITY, Martin Davis. Classic graduate-level introduction to theory of computability, usually referred to as theory of recurrent functions. New preface and appendix. 288pp. 5⅜ × 8½. 61471-9 Pa. $6.95

GENERAL CHEMISTRY, Linus Pauling. Revised 3rd edition of classic first-year text by Nobel laureate. Atomic and molecular structure, quantum mechanics, statistical mechanics, thermodynamics correlated with descriptive chemistry. Problems. 992pp. 5⅜ × 8½. 65622-5 Pa. $18.95

AN INTRODUCTION TO MATRICES, SETS AND GROUPS FOR SCIENCE STUDENTS, G. Stephenson. Concise, readable text introduces sets, groups, and most importantly, matrices to undergraduate students of physics, chemistry, and engineering. Problems. 164pp. 5⅜ × 8½. 65077-4 Pa. $5.95

THE HISTORICAL BACKGROUND OF CHEMISTRY, Henry M. Leicester. Evolution of ideas, not individual biography. Concentrates on formulation of a coherent set of chemical laws. 260pp. 5⅜ × 8½. 61053-5 Pa. $6.00

THE PHILOSOPHY OF MATHEMATICS: An Introductory Essay, Stephan Körner. Surveys the views of Plato, Aristotle, Leibniz & Kant concerning propositions and theories of applied and pure mathematics. Introduction. Two appendices. Index. 198pp. 5⅜ × 8½. 25048-2 Pa. $5.95

THE DEVELOPMENT OF MODERN CHEMISTRY, Aaron J. Ihde. Authoritative history of chemistry from ancient Greek theory to 20th-century innovation. Covers major chemists and their discoveries. 209 illustrations. 14 tables. Bibliographies. Indices. Appendices. 851pp. 5⅜ × 8½. 64235-6 Pa. $15.95

THE FOUR-COLOR PROBLEM: Assaults and Conquest, Thomas L. Saaty and Paul G. Kainen. Engrossing, comprehensive account of the century-old combinatorial topological problem, its history and solution. Bibliographies. Index. 110 figures. 228pp. 5⅜ × 8½.   65092-8 Pa. $6.00

CATALYSIS IN CHEMISTRY AND ENZYMOLOGY, William P. Jencks. Exceptionally clear coverage of mechanisms for catalysis, forces in aqueous solution, carbonyl- and acyl-group reactions, practical kinetics, more. 864pp. 5⅜ × 8½.   65460-5 Pa. $18.95

PROBABILITY: An Introduction, Samuel Goldberg. Excellent basic text covers set theory, probability theory for finite sample spaces, binomial theorem, much more. 360 problems. Bibliographies. 322pp. 5⅜ × 8½.   65252-1 Pa. $7.95

LIGHTNING, Martin A. Uman. Revised, updated edition of classic work on the physics of lightning. Phenomena, terminology, measurement, photography, spectroscopy, thunder, more. Reviews recent research. Bibliography. Indices. 320pp. 5⅜ × 8¼.   64575-4 Pa. $7.95

PROBABILITY THEORY: A Concise Course, Y.A. Rozanov. Highly readable, self-contained introduction covers combination of events, dependent events, Bernoulli trials, etc. Translation by Richard Silverman. 148pp. 5⅜ × 8¼.
                                                                                          63544-9 Pa. $4.50

THE CEASELESS WIND: An Introduction to the Theory of Atmospheric Motion, John A. Dutton. Acclaimed text integrates disciplines of mathematics and physics for full understanding of dynamics of atmospheric motion. Over 400 problems. Index. 97 illustrations. 640pp. 6 × 9.   65096-0 Pa. $16.95

STATISTICS MANUAL, Edwin L. Crow, et al. Comprehensive, practical collection of classical and modern methods prepared by U.S. Naval Ordnance Test Station. Stress on use. Basics of statistics assumed. 288pp. 5⅜ × 8½.
                                                                                          60599-X Pa. $6.00

WIND WAVES: Their Generation and Propagation on the Ocean Surface, Blair Kinsman. Classic of oceanography offers detailed discussion of stochastic processes and power spectral analysis that revolutionized ocean wave theory. Rigorous, lucid. 676pp. 5⅜ × 8½.   64652-1 Pa. $14.95

STATISTICAL METHOD FROM THE VIEWPOINT OF QUALITY CONTROL, Walter A. Shewhart. Important text explains regulation of variables, uses of statistical control to achieve quality control in industry, agriculture, other areas. 192pp. 5⅜ × 8½.   65232-7 Pa. $6.00

THE INTERPRETATION OF GEOLOGICAL PHASE DIAGRAMS, Ernest G. Ehlers. Clear, concise text emphasizes diagrams of systems under fluid or containing pressure; also coverage of complex binary systems, hydrothermal melting, more. 288pp. 6½ × 9¼.   65389-7 Pa. $8.95

STATISTICAL ADJUSTMENT OF DATA, W. Edwards Deming. Introduction to basic concepts of statistics, curve fitting, least squares solution, conditions without parameter, conditions containing parameters. 26 exercises worked out. 271pp. 5⅜ × 8½.   64685-8 Pa. $7.95

DE RE METALLICA, Georgius Agricola. The famous Hoover translation of greatest treatise on technological chemistry, engineering, geology, mining of early modern times (1556). All 289 original woodcuts. 638pp. 6¾ × 11.
60006-8 Clothbd. $15.95

SOME THEORY OF SAMPLING, William Edwards Deming. Analysis of the problems, theory and design of sampling techniques for social scientists, industrial managers and others who find statistics increasingly important in their work. 61 tables. 90 figures. xvii + 602pp. 5⅜ × 8½.
64684-X Pa. $14.95

THE VARIOUS AND INGENIOUS MACHINES OF AGOSTINO RAMELLI: A Classic Sixteenth-Century Illustrated Treatise on Technology, Agostino Ramelli. One of the most widely known and copied works on machinery in the 16th century. 194 detailed plates of water pumps, grain mills, cranes, more. 608pp. 9 × 12.
25497-6 Clothbd. $34.95

LINEAR PROGRAMMING AND ECONOMIC ANALYSIS, Robert Dorfman, Paul A. Samuelson and Robert M. Solow. First comprehensive treatment of linear programming in standard economic analysis. Game theory, modern welfare economics, Leontief input-output, more. 525pp. 5⅜ × 8½.
65491-5 Pa. $12.95

ELEMENTARY DECISION THEORY, Herman Chernoff and Lincoln E. Moses. Clear introduction to statistics and statistical theory covers data processing, probability and random variables, testing hypotheses, much more. Exercises. 364pp. 5⅜ × 8½.
65218-1 Pa. $8.95

THE COMPLEAT STRATEGYST: Being a Primer on the Theory of Games of Strategy, J.D. Williams. Highly entertaining classic describes, with many illustrated examples, how to select best strategies in conflict situations. Prefaces. Appendices. 268pp. 5⅜ × 8½.
25101-2 Pa. $5.95

MATHEMATICAL METHODS OF OPERATIONS RESEARCH, Thomas L. Saaty. Classic graduate-level text covers historical background, classical methods of forming models, optimization, game theory, probability, queueing theory, much more. Exercises. Bibliography. 448pp. 5⅜ × 8¼.
65703-5 Pa. $12.95

CONSTRUCTIONS AND COMBINATORIAL PROBLEMS IN DESIGN OF EXPERIMENTS, Damaraju Raghavarao. In-depth reference work examines orthogonal Latin squares, incomplete block designs, tactical configuration, partial geometry, much more. Abundant explanations, examples. 416pp. 5⅜ × 8¼.
65685-3 Pa. $10.95

THE ABSOLUTE DIFFERENTIAL CALCULUS (CALCULUS OF TENSORS), Tullio Levi-Civita. Great 20th-century mathematician's classic work on material necessary for mathematical grasp of theory of relativity. 452pp. 5⅜ × 8½.
63401-9 Pa. $9.95

VECTOR AND TENSOR ANALYSIS WITH APPLICATIONS, A.I. Borisenko and I.E. Tarapov. Concise introduction. Worked-out problems, solutions, exercises. 257pp. 5⅜ × 8¼.
63833-2 Pa. $6.95

TENSOR CALCULUS, J.L. Synge and A. Schild. Widely used introductory text covers spaces and tensors, basic operations in Riemannian space, non-Riemannian spaces, etc. 324pp. 5⅜ × 8¼. 63612-7 Pa. $7.00

A CONCISE HISTORY OF MATHEMATICS, Dirk J. Struik. The best brief history of mathematics. Stresses origins and covers every major figure from ancient Near East to 19th century. 41 illustrations. 195pp. 5⅜ × 8½. 60255-9 Pa. $7.95

A SHORT ACCOUNT OF THE HISTORY OF MATHEMATICS, W.W. Rouse Ball. One of clearest, most authoritative surveys from the Egyptians and Phoenicians through 19th-century figures such as Grassman, Galois, Riemann. Fourth edition. 522pp. 5⅜ × 8½. 20630-0 Pa. $9.95

HISTORY OF MATHEMATICS, David E. Smith. Non-technical survey from ancient Greece and Orient to late 19th century; evolution of arithmetic, geometry, trigonometry, calculating devices, algebra, the calculus. 362 illustrations. 1,355pp. 5⅜ × 8½. 20429-4, 20430-8 Pa., Two-vol. set $21.90

THE GEOMETRY OF RENÉ DESCARTES, René Descartes. The great work founded analytical geometry. Original French text, Descartes' own diagrams, together with definitive Smith-Latham translation. 244pp. 5⅜ × 8½. 60068-8 Pa. $6.00

THE ORIGINS OF THE INFINITESIMAL CALCULUS, Margaret E. Baron. Only fully detailed and documented account of crucial discipline: origins; development by Galileo, Kepler, Cavalieri; contributions of Newton, Leibniz, more. 304pp. 5⅜ × 8½. (Available in U.S. and Canada only) 65371-4 Pa. $7.95

THE HISTORY OF THE CALCULUS AND ITS CONCEPTUAL DEVELOPMENT, Carl B. Boyer. Origins in antiquity, medieval contributions, work of Newton, Leibniz, rigorous formulation. Treatment is verbal. 346pp. 5⅜ × 8½. 60509-4 Pa. $6.95

THE THIRTEEN BOOKS OF EUCLID'S ELEMENTS, translated with introduction and commentary by Sir Thomas L. Heath. Definitive edition. Textual and linguistic notes, mathematical analysis. 2500 years of critical commentary. Not abridged. 1,414pp. 5⅜ × 8½. 60088-2, 60089-0, 60090-4 Pa., Three-vol. set $26.85

A HISTORY OF VECTOR ANALYSIS: The Evolution of the Idea of a Vectorial System, Michael J. Crowe. The first large-scale study of the history of vector analysis, now the standard on the subject. Unabridged republication of the edition published by University of Notre Dame Press, 1967, with second preface by Michael C. Crowe. Index. 278pp. 5⅜ × 8½. 64955-5 Pa. $7.00

THE HISTORICAL ROOTS OF ELEMENTARY MATHEMATICS, Lucas N.H. Bunt, Phillip S. Jones, and Jack D. Bedient. Fundamental underpinnings of modern arithmetic, algebra, geometry and number systems derived from ancient civilizations. 320pp. 5⅜ × 8½. 25563-8 Pa. $7.95

CALCULUS REFRESHER FOR TECHNICAL PEOPLE, A. Albert Klaf. Covers important aspects of integral and differential calculus via 756 questions. 566 problems, most answered. 431pp. 5⅜ × 8½. 20370-0 Pa. $7.95

CHALLENGING MATHEMATICAL PROBLEMS WITH ELEMENTARY SOLUTIONS, A.M. Yaglom and I.M. Yaglom. Over 170 challenging problems on probability theory, combinatorial analysis, points and lines, topology, convex polygons, many other topics. Solutions. Total of 445pp. 5⅜ × 8½. Two-vol. set.
Vol. I 65536-9 Pa. $5.95
Vol. II 65537-7 Pa. $5.95

FIFTY CHALLENGING PROBLEMS IN PROBABILITY WITH SOLUTIONS, Frederick Mosteller. Remarkable puzzlers, graded in difficulty, illustrate elementary and advanced aspects of probability. Detailed solutions. 88pp. 5⅜ × 8½.
65355-2 Pa. $3.95

EXPERIMENTS IN TOPOLOGY, Stephen Barr. Classic, lively explanation of one of the byways of mathematics. Klein bottles, Moebius strips, projective planes, map coloring, problem of the Koenigsberg bridges, much more, described with clarity and wit. 43 figures. 210pp. 5⅜ × 8½.
25933-1 Pa. $4.95

RELATIVITY IN ILLUSTRATIONS, Jacob T. Schwartz. Clear non-technical treatment makes relativity more accessible than ever before. Over 60 drawings illustrate concepts more clearly than text alone. Only high school geometry needed. Bibliography. 128pp. 6⅛ × 9¼.
25965-X Pa. $5.95

AN INTRODUCTION TO ORDINARY DIFFERENTIAL EQUATIONS, Earl A. Coddington. A thorough and systematic first course in elementary differential equations for undergraduates in mathematics and science, with many exercises and problems (with answers). Index. 304pp. 5⅜ × 8¼.
65942-9 Pa. $7.95

FOURIER SERIES AND ORTHOGONAL FUNCTIONS, Harry F. Davis. An incisive text combining theory and practical example to introduce Fourier series, orthogonal functions and applications of the Fourier method to boundary-value problems. 570 exercises. Answers and notes. 416pp. 5⅜ × 8½.
65973-9 Pa. $8.95

THE THOERY OF BRANCHING PROCESSES, Theodore E. Harris. First systematic, comprehensive treatment of branching (i.e. multiplicative) processes and their applications. Galton-Watson model, Markov branching processes, electron-photon cascade, many other topics. Rigorous proofs. Bibliography. 240pp. 5⅜ × 8½.
65952-6 Pa. $6.95

AN INTRODUCTION TO ALGEBRAIC STRUCTURES, Joseph Landin. Superb self-contained text covers "abstract algebra": sets and numbers, theory of groups, theory of rings, much more. Numerous well-chosen examples, exercises. 247pp. 5⅜ × 8½.
65940-2 Pa. $6.95

GAMES AND DECISIONS: Introduction and Critical Survey, R. Duncan Luce and Howard Raiffa. Superb non-technical introduction to game theory, primarily applied to social sciences. Utility theory, zero-sum games, n-person games, decision-making, much more. Bibliography. 509pp. 5⅜ × 8½. 65943-7 Pa. $10.95